Pearson Edexcel

Level 2

Extended Mathematics Certificate

Diane Oliver, Katherine Pate, Dave Swan
Samantha Burns, Jenny Moulds

Published by Pearson Education Limited, 80 Strand, London, WC2R 0RL.

www.pearsonschoolsandfecolleges.co.uk

Copies of official specifications for all Pearson qualifications may be found on the website: qualifications.pearson.com

Text © Pearson Education Limited 2024
Project managed, and edited by Integra Software Services Pvt. Ltd
Designed and typeset by EMC Design Ltd
Original illustrations by D'Avilla Illustration © Pearson Education Limited 2024
Cover design by EMC Design Ltd
Cover photo © Adrienne Bresnahan/Getty Images

The rights of Diane Oliver, Katherine Pate, Dave Swan, Samantha Burns and Jenny Moulds to be identified as authors of this work have been asserted by them in accordance with the Copyright, Designs and Patents Act 1988.

This publication is protected by copyright, and permission should be obtained from the publisher prior to any prohibited reproduction, storage in a retrieval system, or transmission in any form or by any means, electronic, mechanical, photocopying, recording, or otherwise.
For information regarding permissions, request forms and the appropriate contacts, please visit
https://www.pearson.com/us/contact-us/permissions.html
Pearson Education Limited Rights and Permissions Department.

Unless otherwise indicated herein, any third party trademarks that may appear in this work are the property of their respective owners and any references to third party trademarks, logos or other trade dress are for demonstrative or descriptive purposes only. Such references are not intended to imply any sponsorship, endorsement, authorisation, or promotion of Pearson Education Limited products by the owners of such marks, or any relationship between the owner and Pearson Education Limited or its affiliates, authors, licensees or distributors.

First published 2024

27 26 25 24
10 9 8 7 6 5 4 3 2 1

British Library Cataloguing in Publication Data
A catalogue record for this book is available from the British Library

ISBN 978 1 292 73811 6

Copyright notice
All rights reserved. No part of this publication may be reproduced in any form or by any means (including photocopying or storing it in any medium by electronic means and whether or not transiently or incidentally to some other use of this publication) without the written permission of the copyright owner, except in accordance with the provisions of the Copyright, Designs and Patents Act 1988 or under the terms of a licence issued by the Copyright Licensing Agency, 5th Floor, Shackleton House, 4 Battle Bridge Lane, London, SE1 2HX (www.cla.co.uk). Applications for the copyright owner's written permission should be addressed to the publisher.

Printed in the Glasgow, United Kingdom by Bell & Bain Limited

Picture Credits
The author and publisher would like to thank the following individuals and organizations for permission to reproduce photographs and texts.

Alamy images: GL Archive 103, Heritage Image Partnership Ltd 148; **Shutterstock:** claudio zaccherini 1, Frontpage 9, J.D.S 28, StockSmartStart 53, Mihai-Bogdan Lazar 83, Rahmo 125, Fineart1 136, Diego Cervo 159.

Contents

Introduction		**iv**
Chapter 1 Number		**1**
	1.1 Laws of indices	2
	1.2 Surds	4
	1.3 Rationalising denominators	6
Chapter 2 Algebraic manipulation		**9**
	2.1 Algebraic indices	10
	2.2 Expanding brackets	12
	2.3 Factorising	17
	2.4 Completing the square	21
	2.5 Algebraic fractions	23
Chapter 3 Graphs		**28**
	3.1 Linear graphs	29
	3.2 Quadratic graphs	33
	3.3 Cubic and quartic graphs	40
	3.4 Trigonometric graphs	48
Chapter 4 More graphs		**53**
	4.1 Translating and reflecting graphs	54
	4.2 Stretching graphs	62
	4.3 Circles	68
	4.4 Exponential and reciprocal graphs	72
	4.5 Non-linear graphs	77
Chapter 5 Functions		**83**
	5.1 Functions	84
	5.2 Composite functions	91
	5.3 Inverse functions	95
	5.4 Transforming functions	98
Chapter 6 Equations and inequalities		**103**
	6.1 Solve equations	104
	6.2 Solve quadratic equations	107
	6.3 Solve simultaneous equations	111
	6.4 Solve inequalities	117
Chapter 7 Pythagoras and trigonometry		**125**
	7.1 Pythagoras and trigonometry	126
	7.2 Sine and cosine rules, and area of a triangle	131
Chapter 8 Probability		**136**
	8.1 The language of probability	137
	8.2 Probability problems	144
Chapter 9 Proof		**148**
	9.1 Proof by deduction	149
	9.2 Proof by exhaustion, and disproof by counter example	153
	9.3 Geometric proof	155
Chapter 10 Vectors		**159**
	10.1 Position vectors	160
	10.2 Solving geometric problems	164
Mixed Practice		**169**
Answers		**176**
Index		**204**

Introduction

Resources to motivate students to achieve their potential

Pearson Edexcel Extended Maths Certificate is a Level 2 Maths qualification. It has been designed for students likely to achieve a grade 7 or higher in GCSE Mathematics to stretch and challenge and bridge the transition to Key Stage 5.

Extended Maths Certificate resources have been designed to support students to dive deeper into maths and work independently to boost their mathematical confidence for Level 2 and beyond. You can be confident that the materials fully support the qualification and have all the tools your students need for independent learning and practice. The resources have been designed for independent learning with fully worked solutions, examples, hints and 'talking points'.

Each chapter is introduced by an engaging image with a mathematical link. You will also find a thought-provoking Maths Challenge, which all take a puzzle approach to spark the interest of students (and teachers!).

Chapters are divided into lessons, determined by the Pearson Edexcel scheme of work. Lessons have been written to be 60–70 minutes in duration, for high achieving students who will be able to work through the content effectively and/or select the most appropriate tasks to focus on.

Each lesson explains concepts and is punctuated with lots of practice questions and worked examples. Each lesson has a practice section, which is a series of questions progressing appropriately from Higher GCSE, through key concepts for Extended Maths Certificate and on to application and understanding.

Each lesson includes an exam-style question so that teachers and students can fully understand the exam requirements. The number of marks is always included and you can find exam style mark schemes on the website: pearsonschools.com/extendedmaths

These features are used throughout the book:

Hint

- Tips or hints to help get started on the question or a tip on a part of the question that is a common misconception.

Talking point

- A question or discussion point to encourage students to reflect on their work for students working independently through the content.

Problem-solving

- To indicate when a question is a problem-solving question

Reasoning

- To indicate when a question is a reasoning question

At the end of the book, there is mixed practice with many exam-style questions. This mixed practice brings topics together to consolidate knowledge from all the chapters and provide extra problem-solving with synoptic questions. Mixed practice could be used as a lesson, or as homework, or used for revision.

Answers to all questions are provided in the back of the book*.

Extended Mathematics Certificate is supported by an ActiveHub Essential Teaching subscription. The ActiveHub digital resources include more support for independent learning including:

- video walkthroughs to guide students step by step through example questions with clear, reassuring commentary

- both a 'Practice' and 'Purposeful Practice' worksheet for every lesson to support consolidation

- 'Solution Bank' for each question to demonstrate solutions and give students the confidence to tackle each concept

*Answers should round to 3 s.f. where appropriate, unless otherwise stated.

We hope that the *Extended Maths Certificate* supports your students to dive deeper into maths, boost their mathematical confidence and progress on their maths journey.

Maths challenge

How many ways can you spell Maths? You can move horizontally or vertically.

```
      s
     s h s
    s h t h s
   s h t a t h s
  s h t a M a t h s
   s h t a t h s
    s h t h s
     s h s
      s
```

Archimedes was a mathematician, scientist and inventor from the ancient Greek city of Syracuse, in Sicily, who lived around 250 BCE. He used surds to prove that the value of π must lie between $\frac{223}{71}$ and $\frac{22}{7}$ by calculating the perimeters of polygons fitted to a circle. He also calculated values for $\sqrt{2}$ and $\sqrt{3}$ that are very close to their actual values.

Chapter 1: Number

In this chapter you will:
- use the laws of indices with positive, negative and fractional indices
- simplify and manipulate surds
- rationalise denominators

Prior knowledge
- use negative indices
- use fractional indices
- simplify surds
- rationalise denominators

Chapter 1: Number

1.1 Laws of indices

Use these laws of indices to simplify powers of the same base.

$$x^m \times x^n = x^{m+n}$$

$$x^m \div x^n = x^{m-n}$$

$$(x^m)^n = x^{mn}$$

$$\left(\frac{a}{b}\right)^m = \frac{a^m}{b^m}$$

The nth root of x. → $x^{\frac{1}{n}} = \sqrt[n]{x}$

$$x^{\frac{m}{n}} = \sqrt[n]{x^m}$$

$x^0 = 1$, where $x \neq 0$

$x^{-n} = \frac{1}{x^n}$, where $x \neq 0$

Example

Evaluate:

a $9^{\frac{1}{2}}$ b $64^{\frac{1}{3}}$ c $49^{\frac{3}{2}}$ d $25^{-\frac{3}{2}}$ e $\left(\frac{9}{16}\right)^{-\frac{3}{2}}$

Using $x^{\frac{1}{n}} = \sqrt[n]{x}$

A square root can be positive or negative, as $+ \times + = +$, and $- \times - = +$.

a $9^{\frac{1}{2}} = \sqrt{9}$
$= \pm 3$

This means the cube root of 64.

b $64^{\frac{1}{3}} = \sqrt[3]{64}$
$= 4$

Using $x^{\frac{m}{n}} = \sqrt[n]{x^m}$.
This means the square root of 49, cubed.

c $49^{\frac{3}{2}} = (\sqrt{49})^3$
$= \pm 343$

Using $x^{-n} = \frac{1}{x^n}$

$\sqrt{25} = \pm 5$

d $25^{-\frac{3}{2}} = \frac{1}{25^{\frac{3}{2}}}$
$= \frac{1}{(\sqrt{25})^3}$
$= \frac{1}{(\pm 5)^3}$
$= \pm \frac{1}{125}$

Using $x^{-n} = \frac{1}{x^n}$

e $\left(\frac{9}{16}\right)^{-\frac{3}{2}} = \frac{1}{\left(\frac{9}{16}\right)^{\frac{3}{2}}}$

$= \left(\frac{16}{9}\right)^{\frac{3}{2}}$

$= \frac{(\sqrt{16})^3}{(\sqrt{9})^3}$

$= \frac{(\pm 4)^3}{(\pm 3)^3}$

$= \pm \frac{64}{27}$

Chapter 1: Number

Practice

1 Evaluate

a $49^{\frac{1}{2}}$
b $\left(\frac{1}{81}\right)^{\frac{1}{2}}$
c $\left(\frac{25}{36}\right)^{\frac{1}{2}}$

d $125^{\frac{1}{3}}$
e $\left(\frac{1}{8}\right)^{\frac{1}{3}}$
f $\left(-\frac{27}{64}\right)^{\frac{1}{3}}$

2 Evaluate

a $25^{-\frac{1}{2}}$
b $\left(\frac{1}{16}\right)^{-\frac{1}{2}}$
c $\left(\frac{49}{100}\right)^{-\frac{1}{2}}$

d $8^{-\frac{1}{3}}$
e $\left(\frac{1}{64}\right)^{-\frac{1}{3}}$
f $\left(-\frac{343}{1000}\right)^{-\frac{1}{3}}$

3 Evaluate

a $25^{\frac{3}{2}}$
b $8^{\frac{2}{3}}$
c $81^{-\frac{3}{4}}$

d $\left(\frac{1}{36}\right)^{\frac{3}{2}}$
e $\left(-\frac{8}{27}\right)^{\frac{4}{3}}$
f $\left(\frac{1}{64}\right)^{-\frac{2}{3}}$

4 Work out the value of n.

a $32 = 2^n$
b $\frac{1}{8} = 2^n$
c $\left(\sqrt{5}\right)^3 = 5^n$

5 Write these as single powers of 3.

a $9^{-\frac{1}{2}}$
b $27^{\frac{1}{5}}$
c $81^{\frac{4}{3}}$

6 $5^a \times 125^{\frac{2}{3}} = 25^{\frac{5}{2}}$
Work out the value of a.

Hint for Q6

Use $x^m \times x^n = x^{m+n}$ and $x^{\frac{m}{n}} = \sqrt[n]{x^m}$.

7 $16^{\frac{3}{2}} \times 2^x = 8^{\frac{1}{5}}$
Work out the value of x.

8 $9^{\frac{3}{2}} \times 3^{2t-1} \times \frac{1}{9^2} = 81$
Work out the value of t.

9 Solve $\dfrac{25^{\frac{3x+1}{2}}}{125^{1.5x}} = 5^{10-x}$

Exam-style question

10 $8^{\frac{2}{3}} \times \frac{1}{4^2} \times 2^{3y-4} \times 32^{\frac{3}{5}} = 64$

Find the value of y. **(6 marks)**

11 Problem-solving Solve $-27^{-\frac{2}{3}} \times 9^3 \times (-3)^{2x} = -243$

Talking point

How do you know what the base number is?

1.1 Laws of indices

Chapter 1: Number

1.2 Surds

Talking point

Why are square roots of square numbers, such as $\sqrt{4}$ and $\sqrt{25}$, not surds?

Surds are irrational numbers. A surd is a multiple of $\sqrt{n}$, where n is an integer that is not a square number. For example, $\sqrt{2}$, $\sqrt{3}$, $\sqrt{6}$ and $4\sqrt{5}$ are surds but $\sqrt{4}$ and $\sqrt{25}$ are not surds.

The following rules apply to surds:

- $\sqrt{mn} = \sqrt{m} \times \sqrt{n}$
- $\sqrt{\dfrac{m}{n}} = \dfrac{\sqrt{m}}{\sqrt{n}}$

Example 1

Using the positive values of the surds, simplify

a $\sqrt{45}$
b $\dfrac{\sqrt{12}}{2}$
c $3\sqrt{6} - 2\sqrt{96} + \sqrt{486}$

Find a factor of 45 that is a square number.

Use the rule $\sqrt{ab} = \sqrt{a} \times \sqrt{b}$

$\sqrt{9} = 3$

a $\sqrt{45} = \sqrt{9 \times 5}$
$= \sqrt{9} \times \sqrt{5}$
$= 3\sqrt{5}$

$\sqrt{12} = \sqrt{4} \times \sqrt{3}$

$\sqrt{4} = 2$

Cancel by 2.

b $\dfrac{\sqrt{12}}{2} = \dfrac{\sqrt{4} \times \sqrt{3}}{2}$
$= \dfrac{2 \times \sqrt{3}}{2}$
$= \sqrt{3}$

$\sqrt{6}$ is a common factor.

Work out the square roots $\sqrt{16}$ and $\sqrt{81}$.

$3 - 8 + 9 = 4$

c $3\sqrt{6} - 2\sqrt{96} + \sqrt{486} = 3\sqrt{6} - 2\sqrt{16}\sqrt{6} + \sqrt{81} \times \sqrt{6}$
$= \sqrt{6}(3 - 2\sqrt{16} + \sqrt{81})$
$= \sqrt{6}(3 - 2 \times 4 + 9)$
$= \sqrt{6}(4)$
$= 4\sqrt{6}$

Example 2

Expand and simplify $(3 - \sqrt{5})(7 + \sqrt{5})$

Expand the brackets completely before simplifying.

Collect like terms, $3\sqrt{5} - 7\sqrt{5} = -4\sqrt{5}$

Simplify any roots if possible $\sqrt{25} = 5$

$(3 - \sqrt{5})(7 + \sqrt{5}) = 3(7 + \sqrt{5}) - \sqrt{5}(7 + \sqrt{5})$
$= 21 + 3\sqrt{5} - 7\sqrt{5} - \sqrt{25}$
$= 16 - 4\sqrt{5}$

Chapter 1: Number

Practice

1 Using the positive values of the surds, simplify

    a $\sqrt{20}$ **b** $\sqrt{18}$ **c** $\sqrt{44}$

    d $\sqrt{63}$ **e** $\sqrt{98}$ **f** $\sqrt{700}$

Exam-style question

2 Write $\sqrt{180}$ in the form $a\sqrt{b}$ where a and b are integers. **(2 marks)**

3 Using the positive values of the surds, simplify

    a $\dfrac{\sqrt{18}}{3}$ **b** $\dfrac{\sqrt{75}}{5}$ **c** $\dfrac{\sqrt{48}}{2}$

4 Using the positive values of the surds, simplify

    a $\sqrt{12} + \sqrt{75}$ **b** $\sqrt{192} - \sqrt{48}$

    c $\sqrt{500} + \sqrt{20} - \sqrt{45}$ **d** $\sqrt{50} - \sqrt{18} + 2\sqrt{32}$

    e $2\sqrt{75} - 2\sqrt{12} + \sqrt{147}$ **f** $3\sqrt{20} + 2\sqrt{125} - 4\sqrt{180}$

5 Expand and simplify if possible

    a $\sqrt{5}(2 - \sqrt{3})$ **b** $\sqrt{3}(4 - \sqrt{5})$ **c** $\sqrt{2}(7 + \sqrt{2})$

    d $(4 + \sqrt{2})(5 - \sqrt{3})$ **e** $(4 + \sqrt{5})(2 - \sqrt{3})$ **f** $(2 + \sqrt{3})(4 + \sqrt{3})$

    g $(2 - \sqrt{2})(3 - \sqrt{2})$ **h** $(3 - \sqrt{7})(2 + \sqrt{7})$ **i** $(4 - \sqrt{5})^2$

    j $(\sqrt{2} + \sqrt{3})(\sqrt{2} - \sqrt{3})$ **k** $(\sqrt{3} + \sqrt{5})^2$ **l** $(\sqrt{5} - \sqrt{2})^2$

Hint for Q5a
$\sqrt{5}(2 - \sqrt{3}) = \sqrt{5} \times 2 - \sqrt{5} \times \sqrt{3}$

Hint for Q5i
$(4 - \sqrt{5})^2 = (4 - \sqrt{5})(4 - \sqrt{5})$

Exam-style question

6 a Find the value of $(3\sqrt{8})^2$. **(2 marks)**

    b Simplify $(\sqrt{5}) + (\sqrt{5})^2 + (\sqrt{5})^3 + (\sqrt{5})^4$. **(3 marks)**

7 Reasoning Ethan writes:

$(3 + \sqrt{5})^2 = 9 + 5 = 14$

Ethan is incorrect. Explain why.

8 Show that $(\sqrt{7} - \sqrt{5})(\sqrt{7} + \sqrt{5}) = 2$.

💬 **Talking point**
When the brackets for **Q5d–i** are expanded, some of the expressions can be simplified and some cannot. How do you know when they can be simplified? Explain why.

1.2 Surds

Chapter 1: Number

1.3 Rationalising denominators

It is not good practice to have a fraction with a surd in the denominator. The process of finding an equivalent fraction with a rational denominator is called rationalising the denominator.

The rules for rationalising denominators are:

- For fractions of the form $\dfrac{a}{\sqrt{b}}$, multiply the numerator and denominator by $\sqrt{b}$.
- For fractions of the form $\dfrac{1}{a + \sqrt{b}}$, multiply the numerator and denominator by $a - \sqrt{b}$.
- For fractions of the form $\dfrac{1}{a - \sqrt{b}}$, multiply the numerator and denominator by $a + \sqrt{b}$.

> **Talking point**
> When rationalising the denominator, why do you need to multiply the numerator and denominator by the same surd?

Example

Rationalise each denominator.

a $\dfrac{1}{\sqrt{7}}$ b $\dfrac{1}{2 - \sqrt{5}}$ c $\dfrac{\sqrt{3} + \sqrt{7}}{\sqrt{3} - \sqrt{7}}$ d $\dfrac{1}{(1 + \sqrt{5})^2}$ e $\dfrac{2 + \sqrt{3}}{\sqrt{8} + \sqrt{18}}$

Multiply the numerator and denominator by $\sqrt{7}$.
$\sqrt{7} \times \sqrt{7} = (\sqrt{7})^2 = 7$

a $\dfrac{1}{\sqrt{7}} = \dfrac{1 \times \sqrt{7}}{\sqrt{7} \times \sqrt{7}}$

$= \dfrac{\sqrt{7}}{7}$

Multiply the numerator and denominator by $(2 + \sqrt{5})$.
$\sqrt{5} \times \sqrt{5} = 5$
$4 - 5 = -1$, $2\sqrt{5} - 2\sqrt{5} = 0$
Divide each term by -1.

b $\dfrac{1}{2 - \sqrt{5}} = \dfrac{1 \times (2 + \sqrt{5})}{(2 - \sqrt{5}) \times (2 + \sqrt{5})}$

$= \dfrac{2 + \sqrt{5}}{4 + 2\sqrt{5} - 2\sqrt{5} - 5}$

$= \dfrac{2 + \sqrt{5}}{-1}$

$= -2 - \sqrt{5}$

Multiply the numerator and denominator by $(\sqrt{3} + \sqrt{7})$.
$\sqrt{3}\sqrt{7} - \sqrt{3}\sqrt{7} = 0$
$\sqrt{3}\sqrt{7} = \sqrt{21}$
Divide each term by 2.

c $\dfrac{\sqrt{3} + \sqrt{7}}{\sqrt{3} - \sqrt{7}} = \dfrac{(\sqrt{3} + \sqrt{7}) \times (\sqrt{3} + \sqrt{7})}{(\sqrt{3} - \sqrt{7}) \times (\sqrt{3} + \sqrt{7})}$

$= \dfrac{3 + \sqrt{3}\sqrt{7} + \sqrt{3}\sqrt{7} + 7}{3 + \sqrt{3}\sqrt{7} - \sqrt{3}\sqrt{7} - 7}$

$= \dfrac{10 + 2\sqrt{21}}{-4}$

$= \dfrac{5 + \sqrt{21}}{-2}$

d $\dfrac{1}{(1+\sqrt{5})^2} = \dfrac{1}{(1+\sqrt{5})(1+\sqrt{5})}$ — Expand the brackets.

$= \dfrac{1}{1+\sqrt{5}+\sqrt{5}+5}$ — Simplify and collect like terms.

$= \dfrac{1}{6+2\sqrt{5}}$

$= \dfrac{1 \times (6-2\sqrt{5})}{(6+2\sqrt{5})(6-2\sqrt{5})}$ — Multiply the numerator and denominator by $(6-2\sqrt{5})$.

$= \dfrac{6-2\sqrt{5}}{36-12\sqrt{5}+12\sqrt{5}-20}$ — $36-20=16, -12\sqrt{5}+12\sqrt{5}=0$

$= \dfrac{6-2\sqrt{5}}{16} = \dfrac{3-\sqrt{5}}{8}$

e $\dfrac{2+\sqrt{3}}{\sqrt{8}+\sqrt{18}} = \dfrac{2+\sqrt{3}}{2\sqrt{2}+3\sqrt{2}} = \dfrac{2+\sqrt{3}}{5\sqrt{2}}$ — Simplify the denominator before rationalising.

$= \dfrac{(2+\sqrt{3}) \times 5\sqrt{2}}{5\sqrt{2} \times 5\sqrt{2}}$ — Multiply the numerator and denominator by $5\sqrt{2}$.

$= \dfrac{10\sqrt{2}+5\sqrt{2}\sqrt{3}}{50}$ — $5\sqrt{2} \times 5\sqrt{2} = 25 \times 2 = 50$

$= \dfrac{2\sqrt{2}+\sqrt{6}}{10}$

Practice

1 Rationalise the denominators and simplify if possible.

- **a** $\dfrac{1}{\sqrt{3}}$
- **b** $\dfrac{1}{\sqrt{7}}$
- **c** $\dfrac{3}{\sqrt{2}}$
- **d** $\dfrac{2}{\sqrt{2}}$
- **e** $\dfrac{5}{\sqrt{5}}$
- **f** $\dfrac{3}{\sqrt{6}}$
- **g** $\dfrac{\sqrt{2}}{\sqrt{3}}$
- **h** $\dfrac{\sqrt{2}}{\sqrt{5}}$
- **i** $\dfrac{\sqrt{3}}{\sqrt{7}}$

2 Rationalise the denominators and simplify if possible.

- **a** $\dfrac{1}{5+\sqrt{3}}$
- **b** $\dfrac{1}{3-\sqrt{2}}$
- **c** $\dfrac{2}{1-\sqrt{5}}$
- **d** $\dfrac{2}{1+\sqrt{2}}$
- **e** $\dfrac{3}{2-\sqrt{3}}$
- **f** $\dfrac{4}{5+\sqrt{7}}$
- **g** $\dfrac{\sqrt{2}}{1+\sqrt{3}}$
- **h** $\dfrac{\sqrt{3}}{1-\sqrt{3}}$
- **i** $\dfrac{\sqrt{3}}{2-\sqrt{2}}$
- **j** $\dfrac{\sqrt{5}}{3+\sqrt{3}}$
- **k** $\dfrac{\sqrt{5}}{4-\sqrt{5}}$
- **l** $\dfrac{\sqrt{7}}{7-\sqrt{7}}$

Chapter 1: Number

 Talking point

Why is $\dfrac{a - b\sqrt{c}}{-d}$ equivalent to $\dfrac{1}{d}(b\sqrt{c} - a)$?

3 Rationalise the denominators and simplify.

a $\dfrac{1}{\sqrt{2} - \sqrt{3}}$ b $\dfrac{5}{\sqrt{3} + \sqrt{5}}$ c $\dfrac{2\sqrt{3}}{\sqrt{3} - \sqrt{7}}$

d $\dfrac{2 + \sqrt{3}}{\sqrt{3} + \sqrt{5}}$ e $\dfrac{\sqrt{2} + \sqrt{3}}{\sqrt{2} - \sqrt{3}}$ f $\dfrac{\sqrt{7} - \sqrt{11}}{\sqrt{11} + \sqrt{7}}$

4 Rationalise the denominators and simplify.

a $\dfrac{1}{(2 + \sqrt{3})^2}$ b $\dfrac{1}{(2 - \sqrt{5})^2}$ c $\dfrac{3}{(5 - \sqrt{3})^2}$

d $\dfrac{4}{(6 + \sqrt{2})^2}$ e $\dfrac{1}{(3 - \sqrt{2})(2 + \sqrt{2})}$ f $\dfrac{3}{(2 + \sqrt{5})(4 - \sqrt{5})}$

g $\dfrac{\sqrt{3}}{(1 + \sqrt{3})^2}$ h $\dfrac{\sqrt{5}}{(5 - \sqrt{2})^2}$ i $\dfrac{\sqrt{2}}{(5 - \sqrt{2})(1 + \sqrt{2})}$

5   a   Simplify $\sqrt{75} - \sqrt{12}$.

     b   Hence, rationalise the denominator and simplify $\dfrac{4 + \sqrt{5}}{\sqrt{75} - \sqrt{12}}$.

Hint for Q6

Simplify the denominator before rationalising.

6 Rationalise the denominators and simplify.

a $\dfrac{2 + \sqrt{5}}{\sqrt{50} - \sqrt{8}}$ b $\dfrac{3 - \sqrt{3}}{\sqrt{27} + \sqrt{75}}$ c $\dfrac{1 + \sqrt{2}}{\sqrt{80} + \sqrt{45}}$

d $\dfrac{\sqrt{3} - 4}{2\sqrt{32} - \sqrt{18}}$ e $\dfrac{2 + \sqrt{5}}{1 - 2\sqrt{48} + \sqrt{12}}$ f $\dfrac{5 + \sqrt{3}}{2 - 3\sqrt{20} + \sqrt{80}}$

Exam-style question

7 Rationalise the denominator of $\dfrac{\sqrt{7} + 7}{7 - \sqrt{7}}$.

Give your answer in the form $\dfrac{a + \sqrt{7}}{b}$ where a and b are integers.   **(3 marks)**

8 **Reasoning** Show that $\dfrac{2 + \sqrt{3}}{\sqrt{80} - \sqrt{20}} = \dfrac{2\sqrt{5} + \sqrt{15}}{10}$

8

Maths challenge

It takes 1242 digits to number the pages of a book. How many 5s are used?

Muhammad ibn Musa Al-Khwarizmi (often shortened to Al-Khwarizmi) wrote the first book on algebra in the early 9th century, in the city of Baghdad. The book was called Al-Jabr (The Compendious Book on Calculation by Completion and Balancing), which is where we get the term algebra from.

Chapter 2: Algebraic manipulation

In this chapter you will:
- use the laws of indices
- expand two or three expressions
- use Pascal's triangle to expand simple binomial expressions
- divide a polynomial by a linear expression
- use the factor theorem to factorise cubic expressions
- complete the square
- cancel factors in algebraic fractions
- add, subtract, multiply and divide algebraic fractions

Prior knowledge
- know and use the laws of indices
- expand two or three binomial expressions
- use long division
- substitute a value for an unknown in polynomials
- apply the four operations to fractions

Chapter 2: Algebraic manipulation

2.1 Algebraic indices

Use the laws of indices to simplify algebraic expressions.

Example 1

Simplify

a $2x^3 \times 5x^2$ **b** $\dfrac{12t^7}{4t^5}$ **c** $(p^3)^2 \times 4p^2$ **d** $(4n^2)^3 \div 2n^4$

a $2x^3 \times 5x^2 = 2 \times 5 \times x^3 \times x^2$ — *Rewrite the expression with the numbers together and the x terms together.*

$= 10 \times x^{3+2}$

$= 10x^5$ — *Multiply the integers together. $x^3 \times x^2 = x^{3+2}$*

b $\dfrac{12t^7}{4t^5} = \dfrac{12}{4} \times \dfrac{t^7}{t^5}$ — *Rewrite the expression with the numbers together and the t terms together.*

$= 3 \times t^{7-5}$

$= 3t^2$ — *Use the rule $x^m \div x^n = x^{m-n}$ to simplify the indices.*

c $(p^3)^2 \times 4p^2 = p^6 \times 4p^2$ — *Use the rule $(x^m)^n = x^{mn}$ to simplify the indices.*

$= 4 \times p^6 \times p^2$

$= 4p^8$ — *Use the rule $x^m \times x^n = x^{m+n}$ to simplify the indices.*

d $(4n^2)^3 \div 2n^4 = \dfrac{4^3 \times (n^2)^3}{2n^4}$ — *Write as a fraction. Use the rule $(xy)^n = x^n y^n$ to simplify the numerator.*

$= \dfrac{4^3}{2} \times \dfrac{(n^2)^3}{n^4}$

$= \dfrac{64}{2} \times \dfrac{n^6}{n^4}$ — *Use the rule $(x^m)^n = x^{mn}$ to simplify the indices.*

$= 32n^2$ — *Use the rule $x^m \div x^n = x^{m-n}$ to simplify the indices.*

Example 2

Simplify

a $\dfrac{a^3 + a^7}{a^2}$ **b** $\dfrac{15b^7 - 12b^5}{3b^4}$

a $\dfrac{a^3 + a^7}{a^2} = \dfrac{a^3}{a^2} + \dfrac{a^7}{a^2}$ — *Divide each term in the numerator by a^2.*

$= a^{3-2} + a^{7-2}$

$= a + a^5$ — *a^1 is the same as a.*

b $\dfrac{15b^7 - 12b^5}{3b^4} = \dfrac{15b^7}{3b^4} - \dfrac{12b^5}{3b^4}$ — *Divide each term in the numerator by $3b^4$.*

$= 5b^{7-4} - 4b^{5-4}$

$= 5b^3 - 4b$ — *Simplify each fraction:*
$\dfrac{15b^7}{3b^4} = \dfrac{15}{3} \times \dfrac{b^7}{b^4} = 5 \times b^{7-4}$
$\dfrac{12b^5}{3b^4} = \dfrac{12}{3} \times \dfrac{b^5}{b^4} = 4 \times b^{5-4}$

Chapter 2: Algebraic manipulation

Practice

1 Simplify

a $x^4 \times x^5$
b $3a^2 \times 4a^7$
c $3p^{-2} \times p^5$
d $\dfrac{7q^6}{q^3}$
e $\dfrac{10y^5}{2y^2}$
f $(3b^3)^2$
g $(z^2)^5 \times 3z^4$
h $(2m^3)^4 \times 5m^2$
i $(3t^4)^2 \times (2t^3)^3$
j $(2n^2)^4 \div 2n^3$
k $(5k^3)^2 \div k^{-2}$
l $(4c^5)^2 \div 2c^{-1}$
m $(5v^2)^3 \times (2v^4)^2 \times v^{-5}$
n $(3x^5)^2 \times (4x^3)^2 \times x^{-1}$
o $(4y^6)^3 \div (2y^5)^3 \times 2y^{-5}$

Hint for Q1a
Use the rule $x^m \times x^n = x^{m+n}$ to simplify the indices.

Hint for Q1d
Use the rule $x^m \div x^n = x^{m-n}$ to simplify the indices.

2 Simplify

a $\dfrac{x^5 + x^8}{x}$
b $\dfrac{8t^7 + 14t^5}{2t^3}$
c $\dfrac{4y^4 - 5y}{2y}$
d $\dfrac{4b^3 + 3b^8}{3b^2}$
e $\dfrac{12z^2 - 4z^3}{6z^2}$
f $\dfrac{3n^4 - 5n^2}{5n}$

💬 **Talking point**
Why can't you simplify the numerators before dividing by the denominators in Q2?

Exam-style question

3 a $(3x^2)^{-4}$ can be written in the form ax^n.
Find the value of a and the value of n. **(2 marks)**

b $2x^{-3}(6x^3 + 5x^4) = c + dx^n$ for all values of x.
Find the values of c, d and n. **(2 marks)**

4 Simplify

a $y \times y^{\frac{1}{2}}$
b $2a^{\frac{3}{2}} \times 4a^{\frac{5}{2}}$
c $2x \times \sqrt{x}$
d $\dfrac{4b}{\sqrt{b}}$
e $\dfrac{12t^3}{2\sqrt{t}}$
f $\sqrt{25m^3}$

Hint for Q4c
$\sqrt{x} = x^{\frac{1}{2}}$

Exam-style question

5 Simplify $(x^{-4})^{-\frac{1}{2}}$. **(1 mark)**

6 Reasoning Given that $y = \dfrac{1}{27}x^3$ express each of following in the form ax^b, where a and b are constants.

a $y^{\frac{1}{3}}$

b $3y^{-1}$

7 Reasoning Show that $\dfrac{(4 - \sqrt{x})^2}{\sqrt{x}}$ can be written as $16x^{-\frac{1}{2}} - 8 + x^{\frac{1}{2}}$

8 Reasoning Given that $\dfrac{y^{\frac{7}{2}} + 6y^4}{\sqrt{y}}$ can be written in the form $y^a + 6y^b$, work out the values of a and b.

Chapter 2: Algebraic manipulation

2.2 Expanding brackets

Expanding brackets means multiplying the expressions in the brackets.

To multiply two expressions, multiply each term in one expression by each term in the other expression.

> **Example 1**
>
> **Expand each expression and simplify if possible.**
>
> **a** $(x + 4)(x - 6)$ **b** $(x - 5y)(x^2 - 3)$
>
> **c** $(x + 2y)^2$ **d** $(x + 1)(x^2 + 2x + 5)$
>
> **a** $(x + 4)(x - 6)$
>
> $= x^2 - 6x + 4x - 24$
>
> $= x^2 - 2x - 24$
>
> **b** $(x - 5y)(x^2 - 3)$
>
> $= x^3 - 3x - 5x^2y + 15y$
>
> **c** $(x + 2y)^2$
>
> $= (x + 2y)(x + 2y)$
>
> $= x^2 + 2xy + 2xy + 4y^2$
>
> $= x^2 + 4xy + 4y^2$
>
> **d** $(x + 1)(x^2 + 2x + 5)$
>
> $= x^3 + 2x^2 + 5x + x^2 + 2x + 5$
>
> $= x^3 + 3x^2 + 7x + 5$

Annotations:
- Multiply x by $(x - 6)$ and then multiply 4 by $(x - 6)$.
- Simplify the answer by collecting like terms. $-6x + 4x = -2x$
- $-5y \times x^2 = -5x^2y$. There are no like terms to collect.
- $(x + 2y)^2$ means $(x + 2y)$ multiplied by itself.
- Simplify the answer by collecting like terms.
- Multiply x by $(x^2 + 2x + 5)$ and then multiply 1 by $(x^2 + 2x + 5)$.
- Simplify the answer by collecting like terms.

Chapter 2: Algebraic manipulation

Example 2

Expand each expression and simplify if possible.

a $x(x + 3)(2x - 5)$ **b** $(x - 2)(x + 1)(x + 4)$

a $x(x + 3)(2x - 5)$
$= (x^2 + 3x)(2x - 5)$
$= x^2(2x - 5) + 3x(2x - 5)$
$= 2x^3 - 5x^2 + 6x^2 - 15x$
$= 2x^3 + x^2 - 15x$

> Start by expanding one pair of brackets.
> $x(x + 3) = x^2 + 3x$

> Simplify the answer by collecting like terms.

b $(x - 2)(x + 1)(x + 4)$
$= (x^2 - x - 2)(x + 4)$
$= x^2(x + 4) - x(x + 4) - 2(x + 4)$
$= x^3 + 4x^2 - x^2 - 4x - 2x - 8$
$= x^3 + 3x^2 - 6x - 8$

> Choose two pairs of brackets to expand first. For example,
> $(x - 2)(x + 1) = x^2 + x - 2x - 2$
> $= x^2 - x - 2$

> Multiplying three linear terms always gives a cube term in the final answer. Here that term is x^3.

Use Pascal's triangle to expand expressions in the form $(x + y)^n$, where n is a whole number.

Here are the expansions of $(x + y)^0$, $(x + y)^1$, $(x + y)^2$, $(x + y)^3$ and $(x + y)^4$.

$(x + y)^0 =$ 1
$(x + y)^1 =$ $1x + 1y$
$(x + y)^2 =$ $1x^2 + 2xy + 1y^2$
$(x + y)^3 =$ $1x^3 + 3x^2y + 3xy^2 + 1y^3$
$(x + y)^4 = 1x^4 + 4x^3y + 6x^2y^2 + 4xy^3 + 1y^4$

> **Talking point**
> What do you notice about the total of the indices for each term?

Pascal's triangle is formed by adding adjacent pairs of numbers to find the numbers on the next row.

Here are the first five rows of Pascal's triangle:

```
          1
        1   1
      1   2 + 1
    1   3 +3  1
  1   4   6   4   1
```

> **Talking point**
> What do you notice about the coefficients in the expansions of $(x + y)^0$, $(x + y)^1$, $(x + y)^2$, $(x + y)^3$ and $(x + y)^4$ and Pascal's triangle?

The nth row of Pascal's triangle gives the coefficients of the terms in the expansion of $(x + y)^{n-1}$.

The $(n + 1)$th row of Pascal's triangle gives the coefficients of the terms in the expansion of $(x + y)^n$.

Chapter 2: Algebraic manipulation

Example 3

a Use the fourth row of Pascal's triangle, 1, 3, 3, 1, to expand $(a + b)^3$.

b Use the fifth row of Pascal's triangle, 1, 4, 6, 4, 1, to expand $(a - 3b)^4$.

a $(a + b)^3 = 1a^3 + 3a^2b + 3ab^2 + 1b^3$
$= a^3 + 3a^2b + 3ab^2 + b^3$

> The coefficients are 1, 3, 3, 1.

b $(x + y)^4 = 1x^4 + 4x^3y + 6x^2y^2 + 4xy^3 + 1y^4$

> First expand $(x + y)^4$. The coefficients are 1, 4, 6, 4, 1.

$(a - 3b)^4 = a^4 + 4a^3(-3b) + 6a^2(-3b)^2 + 4a(-3b)^3 + (-3b)^4$

> $(a - 3b)^4$ is the expansion of $(x + y)^4$ with $x = a$ and $y = -3b$.

$= a^4 - 12a^3b + 54a^2b^2 - 108ab^3 + 81b^4$

> Take care with the negative numbers. For example, $4a(-3b)^3 = 4a \times (-27b^3) = -108ab^3$.

Practice

1 Expand each expression and simplify if possible.

a $(x + 4)(x - 3)$ **b** $(x - 7)(x - 8)$ **c** $(x - 6)^2$

d $(x - 3y)^2$ **e** $(x - 5y)(x + 9y)$ **f** $(2x - 3y)(x - 8y)$

g $(5x - 4y)^2$ **h** $(x + 2y)(x^2 - 3)$ **i** $(x + 1)(x^2 + 3x + 2)$

j $(2x + 3)(x + 2y - 1)$ **k** $(3y - 4x + 1)(3x - 2)$ **l** $(x + 3)(x^2 - 4y^2)$

m $(x - 5)(2x^2 + 5x + 7)$ **n** $(x - 2)(3x^2 - x - 4)$ **o** $(x + 4)(5x^2 - x - 2)$

Exam-style question

2 a Expand $(a + b)^2$.   **(1 mark)**

b Hence or otherwise, expand $(2y + 3z)^2$.   **(2 marks)**

Chapter 2: Algebraic manipulation

3   Expand each expression and simplify if possible.

    a $x(x+2)(x+3)$ **b** $x(x+1)(x-4)$ **c** $x(x+4)(2x-1)$

    d $x(2x-3)(3x-4)$ **e** $2x(x-1)(4x+3)$ **f** $5x(2x+1)(3x-4)$

    g $x(x+1)(x-y+3)$ **h** $3x(x-2)(2x+3y-1)$ **i** $x(x+3y)(2x-y+4)$

    j $(x+1)(x+2)(x+3)$ **k** $(x-1)(x+2)(x+3)$ **l** $(x-1)(x+2)(x+4)$

    m $(x-1)(x-2)(x+4)$ **n** $(x-1)(x-3)(x-4)$ **o** $(x-2)(x+3)(x-5)$

    p $(2x+1)(x+3)(x-4)$ **q** $(2x-3)(3x-1)(x-2)$ **r** $(x+4)(3x-1)(5x+2)$

    s $(x+y)(x+1)(x+2)$ **t** $(x+y)(x+y)(x-7)$ **u** $(3x+2y)^3$

4 **a** Use the fourth row of Pascal's triangle, 1, 3, 3, 1, to expand $(c+d)^3$.

    b Use the fifth row of Pascal's triangle, 1, 4, 6, 4, 1, to expand $(a+b)^4$.

    c Use the sixth row of Pascal's triangle, 1, 5, 10, 10, 5, 1, to expand $(p+q)^5$.

5 **a** Use the fourth row of Pascal's triangle, 1, 3, 3, 1, to expand

      i $(x+2y)^3$ **ii** $(x-2y)^3$ **iii** $(2x-3y)^3$

    b Use the fifth row of Pascal's triangle, 1, 4, 6, 4, 1, to expand

      i $(x+2y)^4$ **ii** $(x-2y)^4$ **iii** $(2x-3y)^4$

6 **a** Use the fourth row of Pascal's triangle, 1, 3, 3, 1, to expand

      i $(x+2)^3$ **ii** $(x-2)^3$ **iii** $(2x-3)^3$

    b Use the fifth row of Pascal's triangle, 1, 4, 6, 4, 1, to expand

      i $(x+2)^4$ **ii** $(x-2)^4$ **iii** $(2x-3)^4$

    c Use the sixth row of Pascal's triangle, 1, 5, 10, 10, 5, 1, to expand $(2x-1)^5$.

7 **a** Use the fifth row of Pascal's triangle, 1, 4, 6, 4, 1, to expand $(1+10x)^4$.

    b Use your expansion, with an appropriate value of x, to find the value of 1001^4.

Hint for Q6ai

$(x+2)^3$ is the expansion of $(x+y)^3$ with $y=2$.

 Talking point

Explain where the minus signs go in the answer when there is a minus sign in the question.

Hint for Q7b

$1001^4 = (1 + 10 \times 100)^4$, therefore, substitute $x = \ldots$

Chapter 2: Algebraic manipulation

Exam-style question

8 Here are the first six rows of Pascal's triangle.

$$\begin{array}{ccccccccccc}
&&&&&1&&&&&\\
&&&&1&&1&&&&\\
&&&1&&2&&1&&&\\
&&1&&3&&3&&1&&\\
&1&&4&&6&&4&&1&\\
1&&5&&10&&10&&5&&1
\end{array}$$

Use this information to expand $(p+q)^4$. (2 marks)

9 Use the fifth row of Pascal's triangle, 1, 4, 6, 4, 1, to expand $\left(1+\frac{x}{2}\right)^4$.

Exam-style question

10 Here are the first six rows of Pascal's triangle.

$$\begin{array}{ccccccccccc}
&&&&&1&&&&&\\
&&&&1&&1&&&&\\
&&&1&&2&&1&&&\\
&&1&&3&&3&&1&&\\
&1&&4&&6&&4&&1&\\
1&&5&&10&&10&&5&&1
\end{array}$$

Use this information to expand and simplify $(2a+b)^3$. (3 marks)

11 **Reasoning** Here are the first six rows of Pascal's triangle.

$$\begin{array}{ccccccccccc}
&&&&&1&&&&&\\
&&&&1&&1&&&&\\
&&&1&&2&&1&&&\\
&&1&&3&&3&&1&&\\
&1&&4&&6&&4&&1&\\
1&&5&&10&&10&&5&&1
\end{array}$$

a Expand and simplify $(x+y)^4 + (x-y)^4$.

b Expand and simplify $(2x+y)^5 + (x-2y)^5$.

Chapter 2: Algebraic manipulation

2.3 Factorising

Factorising is the opposite of expanding brackets.

Factorising an expression means writing it as a product of its factors.

For example, $\frac{2}{3}x^2 + \frac{10}{3}x = \frac{2}{3}x(x + 5)$ because $\frac{2}{3}x$ is a common factor of $\frac{2}{3}x^2$ and $\frac{10}{3}x$.

A quadratic expression has the form $ax^2 + bx + c$ where $a \neq 0$. To factorise a quadratic expression:

- find two factors of ac that total b
- rewrite the b term as a sum of the two factors
- factorise each pair of terms
- take out the common factor.

A quadratic expression of the form $x^2 - y^2$ is called the 'difference of two squares' and factorises to $(x + y)(x - y)$.

Example 1

Factorise

a $x^2 - 9x + 20$ **b** $6x^2 + 7x - 20$ **c** $9x^2 - 64y^2$

a $x^2 - 9x + 20$ — $a = 1, b = -9$ and $c = 20$

$ac = 20$ and $b = -9$

So $x^2 - 9x + 20 = x^2 - 5x - 4x + 20$ — -4 and -5 are factors of 20 that add to -9. Rewrite the b term using these two factors.

$= x(x - 5) - 4(x - 5)$ — Factorise the first two terms and the last two terms.

$= (x - 5)(x - 4)$ — $(x - 5)$ is a factor of both terms $x(x - 5)$ and $-4(x - 5)$. To completely factorise, take $(x - 5)$ outside the brackets.

b $6x^2 + 7x - 20$ — $a = 6, b = 7$ and $c = -20$

$ac = -120$ and $b = 7$

So $6x^2 + 7x - 20 = 6x^2 + 15x - 8x - 20$ — 15 and -8 are factors of -120 that add to 7. Rewrite the b term using these two factors.

$= 3x(2x + 5) - 4(2x + 5)$

$= (2x + 5)(3x - 4)$ — Factorise the first two terms and the last two terms.

$(2x + 5)$ is a factor of both terms $3x(2x + 5)$ and $-4(2x + 5)$. To completely factorise, take $(2x + 5)$ outside the brackets.

c $9x^2 - 64y^2 = 3^2x^2 - 8^2y^2$

$= (3x + 8y)(3x - 8y)$ — This is the difference of two squares as the two terms are $(3x)^2$ and $(8y)^2$.

Chapter 2: Algebraic manipulation

 Talking point

When expanding and simplifying the brackets, why are there only two terms?

Example 2

Factorise completely

a $5x^2 - x^3$
b $x^3 - 6x^2 + 5x$

x^2 is a common factor of both terms.
This does not factorise any further.

a $5x^2 - x^3 = x^2(5 - x)$

x is a common factor of all three terms.

b $x^3 - 6x^2 + 5x = x(x^2 - 6x + 5)$

Factorise the quadratic expression $x^2 - 6x + 5$.

$= x(x - 5)(x - 1)$

Use the factor theorem to find simple linear factors of a polynomial.

The factor theorem states that, if a polynomial is equal to 0 for any value p, then $(x - p)$ is a factor of the polynomial.

To factorise a polynomial, use long division to divide by $(x \pm p)$, where p is an integer.

Example 3

Show that $(x - 4)$ is a factor of $x^3 - x^2 - 10x - 8$ by using

a the factor theorem
b algebraic division.

Substitute $p = 4$ into the polynomial to show it is equal to 0.

a For $(x - 4)$, $p = 4$
$4^3 - 4^2 - 10 \times 4 - 8 = 64 - 16 - 40 - 8 = 0$

Here, $p = 4$ and $x^3 - x^2 - 10x - 8 = 0$ when $p = 4$, so $(x - 4)$ is a factor of $x^3 - x^2 - 10x - 8$.

So $(x - 4)$ is a factor of $x^3 - x^2 - 10x - 8$.

Start by dividing the first term of the polynomial by x.
$x^3 \div x = x^2$

b $\quad\quad\quad x^2 + 3x + 2$
$x - 4 \overline{)x^3 - x^2 - 10x - 8}$
$\quad\quad x^3 - 4x^2$
$\quad\quad\quad\quad 3x^2 - 10x$
$\quad\quad\quad\quad 3x^2 - 12x$
$\quad\quad\quad\quad\quad\quad 2x - 8$
$\quad\quad\quad\quad\quad\quad 2x - 8$
$\quad\quad\quad\quad\quad\quad\quad\quad 0$

Copy $-10x$.

Multiply $(x - 4)$ by x^2.
$x^2 \times (x - 4) = x^3 - 4x^2$

Repeat the method.
Divide $3x^2$ by x. $3x^2 \div x = 3x$
Multiply $(x - 4)$ by $3x$.
$3x \times (x - 4) = 3x^2 - 12x$

Subtract.
$(x^3 - x^2) - (x^3 - 4x^2) = 3x^2$

Subtract.
$(3x^2 - 10x) - (3x^2 - 12x) = 2x$

Copy -8.

So $(x - 4)$ is a factor of $x^3 - x^2 - 10x - 8$.

Repeat the method.
Divide $2x$ by x. $2x \div x = 2$
Multiply $(x - 4)$ by 2.
$2 \times (x - 4) = 2x - 8$

Subtract. $(2x - 8) - (2x - 8) = 0$
No numbers are left to copy so the division is finished.
The remainder is 0, so $(x - 4)$ is a factor of $x^3 - x^2 - 10x - 8$.

Chapter 2: Algebraic manipulation

Example 4

Given that $(x + 3)$ is a factor of $x^3 - x^2 - 9x + 9$, fully factorise $x^3 - x^2 - 9x + 9$.

$$\begin{array}{r}
x^2 - 4x + 3 \\
x+3{\overline{\smash{\big)}\,x^3 - x^2 - 9x + 9}} \\
\underline{x^3 + 3x^2} \\
-4x^2 - 9x \\
\underline{-4x^2 - 12x} \\
3x + 9 \\
\underline{3x + 9} \\
0
\end{array}$$

$x^3 - x^2 - 9x + 9 = (x + 3)(x^2 - 4x + 3)$ — Divide $x^3 - x^2 - 9x + 9$ by $(x + 3)$.

$ = (x + 3)(x - 1)(x - 3)$ — Factorise $x^2 - 4x + 3$.

Practice

1 Factorise completely

- **a** $10x - 25$
- **b** $18x^2 + 12$
- **c** $\frac{1}{2}x^2 - \frac{3}{2}x$
- **d** $x^2 - 16$
- **e** $\frac{2}{5}x^2 + \frac{12}{5}x$
- **f** $\frac{6}{7}x^2 + \frac{10}{7}x$
- **g** $\frac{16}{3}x - \frac{20}{3}x^2$
- **h** $\frac{8}{11}x - \frac{12}{11}xy$
- **i** $25 - x^2$
- **j** $x^2 - 25y^2$
- **k** $\frac{81}{100}x^2 - \frac{4}{49}y^2$
- **l** $36 - 4y^2$
- **m** $x^2 + x - 12$
- **n** $x^2 + 4x - 5$
- **o** $x^2 + x - 20$
- **p** $x^2 + x - 30$
- **q** $x^2 - 9x + 14$
- **r** $x^2 - 4x - 21$
- **s** $4x^2 - 7x - 2$
- **t** $3x^2 + 7x - 6$
- **u** $2x^2 + 9x + 4$
- **v** $12x^2 - 11x + 2$
- **w** $10x^2 - 9x + 2$
- **x** $-6x^2 + 11x - 3$

2 Factorise completely

- **a** $x^3 + 3x^2$
- **b** $2x^2 - 7x^3$
- **c** $12x^2 - 20x^3$
- **d** $x^3 + 5x^2 + 6x$
- **e** $x^3 + x^2 - 2x$
- **f** $x^3 - 9x$
- **g** $x^3 + x^2 - 20x$
- **h** $x^3 - 7x^2 + 12x$
- **i** $3x^3 - 14x^2 - 24x$
- **j** $4x^3 + 10x^2 + 6x$
- **k** $12x^3 - 3x$
- **l** $-12x^3 - 34x^2 - 10x$

3 Use the factor theorem to show that

- **a** $(x - 1)$ is a factor of $x^3 - x^2 - 10x + 10$
- **b** $(x + 1)$ is a factor of $x^3 + 2x^2 - 6x - 7$
- **c** $(x + 2)$ is a factor of $x^3 + x^2 - 3x - 2$
- **d** $(x - 3)$ is a factor of $x^3 - 2x^2 - 5x + 6$
- **e** $(x - 5)$ is a factor of $x^3 - 3x^2 - 7x - 15$
- **f** $(x + 4)$ is a factor of $x^3 + 5x^2 - x - 20$

Chapter 2: Algebraic manipulation

Talking point
To show a linear expression is a factor of a polynomial, do you prefer using the factor theorem or algebraic division? Explain why.

Hint for Q7b
Write $x^3 + 3x^2 - 4$ as $x^3 + 3x^2 + 0x - 4$ when dividing it by $(x - 1)$.

4 Repeat **Q3** using algebraic division.

5 **a** **Reasoning** Show that $(x + 3)$ is a factor of $2x^3 + 9x^2 + 4x - 15$.

   b Hence, fully factorise $2x^3 + 9x^2 + 4x - 15$.

6 **a** **Reasoning** Show that $(x + 5)$ is a factor of $x^3 + 4x^2 - 25x - 100$.

   b Hence, fully factorise $x^3 + 4x^2 - 25x - 100$.

7 **a** **Reasoning** Show that $(x - 1)$ is a factor of $x^3 + 3x^2 - 4$.

   b Hence, fully factorise $x^3 + 3x^2 - 4$.

8 Given that $(x + 3)$ is a factor of $6x^3 + 17x^2 - 5x - 6$, fully factorise $6x^3 + 17x^2 - 5x - 6$.

9 Given that $(x + 4)$ is a factor of $2x^3 + 21x^2 + 72x + 80$, fully factorise $2x^3 + 21x^2 + 72x + 80$.

10 **a** **Reasoning** Show that $(x - 4)$ is a factor of $5x^3 - 17x^2 - 5x - 28$.

   b Hence, express $5x^3 - 17x^2 - 5x - 28$ in the form $(x - 4)(Ax^2 + Bx + C)$, where A, B and C are integers.

11 **a** **Reasoning** Show that $4x^3 - 27x + 27 = (x + 3)(px - q)^2$, where p and q are integers.

   b Hence, fully factorise $4x^3 - 27x + 27$.

Exam-style question

12 **a** Use the factor theorem to show that $(x - 2)$ is a factor of $x^3 - 2x^2 - x + 2$. **(2 marks)**

   b Hence, write $x^3 - 2x^2 - x + 2$ in the form $(x - 2)(x + a)(x + b)$, where a and b are integers. **(3 marks)**

Chapter 2: Algebraic manipulation

2.4 Completing the square

Expressions such as $(x + 1)^2$, $(x - 2)^2$ and $\left(x + \frac{1}{2}\right)^2$ are called perfect squares.

Perfect squares form part of the process for completing the square.

To complete the square for a quadratic expression in the form $x^2 + bx + c$, find the perfect square that gives the same x^2 and x terms as the quadratic expression. Then subtract the perfect square's constant term from the constant term given in the quadratic expression.

That is, to complete the square for a quadratic expression in the form $x^2 + bx + c$, rewrite it as

$$\left(x + \frac{b}{2}\right)^2 - \left(\frac{b}{2}\right)^2 + c$$

> **Talking point**
> How can you show that
> $\left(x + \frac{b}{2}\right)^2 - \left(\frac{b}{2}\right)^2 + c = x^2 + bx + c$?

To complete the square for a quadratic equation in the form $ax^2 + bx + c$, first factorise the first two terms using a as the common factor. That is, rewrite it as

$$a\left(x + \frac{b}{2a}\right)^2 - \frac{b^2}{4a} + c$$

> **Talking point**
> How can you show that
> $a\left(x + \frac{b}{2a}\right)^2 - \frac{b^2}{4a} + c = ax^2 + bx + c$?

Example

Complete the square for each expression.

a $x^2 + 4x$ **b** $x^2 + 5x + 3$ **c** $2x^2 + 3x - 1$

a $x^2 + 4x = (x + 2)^2 - 2^2$
$\qquad\quad = (x + 2)^2 - 4$

— Halve the coefficient of x. Write the expression in the form $\left(x + \frac{b}{2}\right)^2 - \left(\frac{b}{2}\right)^2 + c$ when $c = 0$.

b $x^2 + 5x + 3 = \left(x + \frac{5}{2}\right)^2 - \left(\frac{5}{2}\right)^2 + 3$
$\qquad\qquad\quad = \left(x + \frac{5}{2}\right)^2 - \frac{13}{4}$

— Write the expression in the form $\left(x + \frac{b}{2}\right)^2 - \left(\frac{b}{2}\right)^2 + c$.

— $-\left(\frac{5}{2}\right)^2 + 3 = -\frac{25}{4} + \frac{12}{4} = -\frac{13}{4}$

c $2x^2 + 3x - 1 = 2\left[x^2 + \frac{3}{2}x\right] - 1$

— Factorise the first two terms using the coefficient of x^2 as the common factor.

$\qquad\qquad\quad = 2\left[\left(x + \frac{3}{4}\right)^2 - \left(\frac{3}{4}\right)^2\right] - 1$

$\qquad\qquad\quad = 2\left[\left(x + \frac{3}{4}\right)^2 - \frac{9}{16}\right] - 1$

— Complete the square for the expression in the square brackets.

$\qquad\qquad\quad = 2\left(x + \frac{3}{4}\right)^2 - \frac{9}{8} - 1$

$\qquad\qquad\quad = 2\left(x + \frac{3}{4}\right)^2 - \frac{17}{8}$

— Simplify and expand the square brackets.

— Note that this expression is in the form $a\left(x + \frac{b}{2a}\right)^2 - \frac{b^2}{4a} + c$.

Chapter 2: Algebraic manipulation

Practice

1 Write these as perfect squares, $(x + p)^2$.

    a $x^2 + 12x + 36$ b $x^2 - 10x + 25$ c $x^2 + x + \frac{1}{4}$

2 Write these expressions in the form $(x + p)^2 - q$, where p and q are constants.

    a $x^2 + 12x$ b $x^2 - 14x$ c $x^2 + 9x$

3 Write these expressions in the form $(x + p)^2 - q$, where p and q are constants.

    a $x^2 + 5x - 4$ b $x^2 - 11x - 3$ c $x^2 - x + 2$

4 Write these expressions in the form $A(x + B)^2 - C$, where A, B and C are constants.

    a $2x^2 - x - 3$ b $3x^2 + 5x - 2$ c $5x^2 - 4x + 1$

    d $4x^2 + 2x - 3$ e $2x^2 + 7x + 9$ f $3x^2 - x - 5$

5 Write $5x^2 - 2x - 1$ in the form $p(x + q)^2 + r$, where p, q and r are constants.

6 Given that $3x^2 + 7x - 4 = A(x + B)^2 + C$, work out the values of A, B and C.

Exam-style question

7 Write the quadratic expression $9x^2 + 15x + 10$ in the form $(ax + b)^2 + c$ where a is an integer and b and c are fractions.   **(4 marks)**

Chapter 2: Algebraic manipulation

2.5 Algebraic fractions

To simplify algebraic fractions, factorise the numerator and denominator where possible and then cancel any common factors.

> **Example 1**
>
> **Simplify**
>
> a $\dfrac{4(x+3)}{5(x+3)}$ b $\dfrac{x^2-25}{x^2+7x+10}$ c $\dfrac{2x^2-7x-4}{6x^2-x-2}$ d $\dfrac{5x^3-8x^2+3x}{25x^2-9}$
>
> a $\dfrac{4(x+3)}{5(x+3)} = \dfrac{4}{5}$ — Cancel the common factor of $(x+3)$.
>
> b $\dfrac{x^2-25}{x^2+7x+10} = \dfrac{(x+5)(x-5)}{(x+2)(x+5)}$ — Factorise. $x^2-25 = (x+5)(x-5)$ and $x^2+7x+10 = (x+2)(x+5)$
>
> $\quad = \dfrac{x-5}{x+2}$ — Cancel the common factor of $(x+5)$.
>
> c $\dfrac{2x^2-7x-4}{6x^2-x-2} = \dfrac{(2x+1)(x-4)}{(2x+1)(3x-2)}$ — Factorise.
> $2x^2-7x-4 = 2x^2+x-8x-4$
> $= x(2x+1) - 4(2x+1)$
> $= (2x+1)(x-4)$ and
> $6x^2-x-2 = 6x^2+3x-4x-2$
> $= 3x(2x+1) - 2(2x+1)$
> $= (2x+1)(3x-2)$
>
> $\quad = \dfrac{x-4}{3x-2}$ — Cancel the common factor of $(2x+1)$.
>
> d $\dfrac{5x^3-8x^2+3x}{25x^2-9} = \dfrac{x(5x-3)(x-1)}{(5x+3)(5x-3)}$ — Factorise.
> $5x^3-8x^2+3x = x(5x^2-8x+3)$
> $= x(5x^2-5x-3x+3)$
> $= x[5x(x-1) - 3(x-1)]$
> $= x(5x-3)(x-1)$ and
> $25x^2-9 = (5x+3)(5x-3)$
>
> $\quad = \dfrac{x(x-1)}{5x+3}$ — Cancel the common factor of $(5x-3)$.

When multiplying algebraic fractions, cancel common factors in the numerators and denominators before multiplying the fractions together.

Dividing by an algebraic fraction is the same as multiplying by its reciprocal.

Chapter 2: Algebraic manipulation

> **Example 2**
>
> Write as a single fraction in its simplest form.
>
> **a** $\dfrac{x+2}{20} \times \dfrac{5}{6x+12}$
>
> **b** $\dfrac{x^2+2x-8}{x^2+x-6} \times \dfrac{2x^2+7x+3}{x^2-16}$
>
> **c** $\dfrac{6x^3+10x^2-4x}{2x^2+x-6} \div \dfrac{6x^2-2x}{2x^2-x-3}$

Factorise the numerators and denominators.
$6x + 12 = 6(x + 2)$

Cancel the common factors 5 and $(x + 2)$.

a $\dfrac{x+2}{20} \times \dfrac{5}{6x+12} = \dfrac{x+2}{20} \times \dfrac{5}{6(x+2)}$

$= \dfrac{1}{4} \times \dfrac{1}{6}$

$= \dfrac{1}{24}$

Factorise the numerators and denominators.

Cancel the common factors $(x + 4)$, $(x + 3)$ and $(x - 2)$.

b $\dfrac{x^2+2x-8}{x^2+x-6} \times \dfrac{2x^2+7x+3}{x^2-16} = \dfrac{(x+4)(x-2)}{(x+3)(x-2)} \times \dfrac{(x+3)(2x+1)}{(x+4)(x-4)}$

$= \dfrac{1}{1} \times \dfrac{(2x+1)}{(x-4)}$

$= \dfrac{2x+1}{x-4}$

Dividing by $\dfrac{6x^2-2x}{2x^2-x-3}$ is the same as multiplying by its reciprocal $\dfrac{2x^2-x-3}{6x^2-2x}$.

Factorise the numerators and denominators.

Cancel the common factors 2, x, $(3x - 1)$, $(x + 2)$ and $(2x - 3)$.

c $\dfrac{6x^3+10x^2-4x}{2x^2+x-6} \div \dfrac{6x^2-2x}{2x^2-x-3} = \dfrac{6x^3+10x^2-4x}{2x^2+x-6} \times \dfrac{2x^2-x-3}{6x^2-2x}$

$= \dfrac{2x(3x-1)(x+2)}{(x+2)(2x-3)} \times \dfrac{(2x-3)(x+1)}{2x(3x-1)}$

$= \dfrac{1}{1} \times \dfrac{x+1}{1}$

$= x + 1$

When adding or subtracting algebraic fractions, use the same method as for adding or subtracting numerical fractions. That is, find the lowest common denominator and then write both fractions as equivalent fractions with that denominator before adding and subtracting. For example,

$\dfrac{1}{2} + \dfrac{1}{4} = \dfrac{2}{4} + \dfrac{1}{4} = \dfrac{3}{4}$

$\dfrac{x}{2} + \dfrac{x}{4} = \dfrac{2x}{4} + \dfrac{x}{4} = \dfrac{3x}{4}$

or $\dfrac{1}{2x} + \dfrac{1}{4x} = \dfrac{2}{4x} + \dfrac{1}{4x} = \dfrac{3}{4x}$.

Chapter 2: Algebraic manipulation

Example 3

Write as a single fraction in its simplest form.

a $\dfrac{x+3}{4} + \dfrac{2x-1}{5}$

b $\dfrac{3}{x-1} - \dfrac{2}{x+4}$

c $\dfrac{x+1}{x-5} - \dfrac{x+3}{x+2}$

a $\dfrac{x+3}{4} + \dfrac{2x-1}{5} = \dfrac{5(x+3)}{20} + \dfrac{4(2x-1)}{20}$ ⟵ Write each fraction as an equivalent fraction with the common denominator 20.

$= \dfrac{5x+15}{20} + \dfrac{8x-4}{20}$ ⟵ Expand and simplify.

$= \dfrac{13x+11}{20}$

b $\dfrac{3}{x-1} - \dfrac{2}{x+4} = \dfrac{3(x+4)}{(x-1)(x+4)} - \dfrac{2(x-1)}{(x-1)(x+4)}$ ⟵ Write each fraction as an equivalent fraction with the common denominator $(x-1)(x+4)$.

$= \dfrac{3x+12}{(x-1)(x+4)} - \dfrac{2x-2}{(x-1)(x+4)}$ ⟵ Expand and simplify.

$= \dfrac{x+14}{(x-1)(x+4)}$

c $\dfrac{x+1}{x-5} - \dfrac{x+3}{x+2} = \dfrac{(x+1)(x+2)}{(x-5)(x+2)} - \dfrac{(x+3)(x-5)}{(x+2)(x-5)}$ ⟵ Write each fraction as an equivalent fraction with the common denominator $(x-5)(x+2)$.

$= \dfrac{x^2+3x+2}{(x+2)(x-5)} - \dfrac{x^2-2x-15}{(x+2)(x-5)}$ ⟵ Expand and simplify.

$= \dfrac{5x+17}{(x+2)(x-5)}$

Example 4

Write as a single fraction in its simplest form

a $\dfrac{1}{x^2-x-6} + \dfrac{2}{x+2}$

b $\dfrac{2x^2-7x-4}{x^2-2x-8} - \dfrac{x^2+3x}{x^2-9}$

a $\dfrac{1}{x^2-x-6} + \dfrac{2}{x+2} = \dfrac{1}{(x+2)(x-3)} + \dfrac{2}{x+2}$ ⟵ First factorise the quadratic expression.

$= \dfrac{1}{(x+2)(x-3)} + \dfrac{2(x-3)}{(x+2)(x-3)}$ ⟵ Write each fraction as an equivalent fraction with the common denominator $(x+2)(x-3)$.

$= \dfrac{1}{(x+2)(x-3)} + \dfrac{2x-6}{(x+2)(x-3)}$ ⟵ Expand and simplify.

$= \dfrac{2x-5}{(x+2)(x-3)}$

b $\dfrac{2x^2-7x-4}{x^2-2x-8} - \dfrac{x^2+3x}{x^2-9} = \dfrac{(2x+1)(x-4)}{(x+2)(x-4)} - \dfrac{x(x+3)}{(x-3)(x+3)}$ ⟵ First factorise each quadratic expression and simplify by cancelling common factors.

$= \dfrac{2x+1}{x+2} - \dfrac{x}{x-3}$

$= \dfrac{(2x+1)(x-3)}{(x+2)(x-3)} - \dfrac{x(x+2)}{(x-3)(x+2)}$ ⟵ Write each fraction as an equivalent fraction with the common denominator $(x+2)(x-3)$.

$= \dfrac{2x^2-5x-3}{(x+2)(x-3)} - \dfrac{x^2+2x}{(x-3)(x+2)}$ ⟵ Expand and simplify.

$= \dfrac{x^2-3x-3}{(x+2)(x-3)}$

Chapter 2: Algebraic manipulation

Practice

1 Simplify

a $\dfrac{x(x-2)}{2(x-2)}$ b $\dfrac{3(x+7)}{(x+7)}$ c $\dfrac{x-3}{(x-3)(x+4)}$

d $\dfrac{x+4}{x^2-16}$ e $\dfrac{x^2-9}{x^2+5x+6}$ f $\dfrac{x^2+7x+6}{x^2+3x-18}$

g $\dfrac{x^2+4x-21}{x^2-49}$ h $\dfrac{4x^2-1}{2x^2+5x-3}$ i $\dfrac{6x^2-9x}{2x^2+7x-15}$

j $\dfrac{2x^2+7x+5}{2x^2-x-3}$ k $\dfrac{6x^2-13x-5}{6x^2-x-1}$ l $\dfrac{x^3-x^2-20x}{x^2-25}$

m $\dfrac{2x^3+7x^2+3x}{x^2+6x+9}$ n $\dfrac{4x^3+8x^2-5x}{6x^2-x-1}$ o $\dfrac{6x^3+15x^2-36x}{2x^3+3x^2-9x}$

> ### Exam-style question
>
> **2** Simplify $\dfrac{(x+3)^2}{x^2+5x+6}$ (2 marks)

3 Reasoning

Hint for Q3a
Use the factor theorem or algebraic division.

a Show that $(x+1)$ is a factor of $x^3 - x^2 - 17x - 15$.

b Hence, fully factorise $x^3 - x^2 - 17x - 15$.

c Hence, simplify $\dfrac{x^3 - x^2 - 17x - 15}{2x^3 - 4x^2 - 30x}$.

4 Reasoning

a Show that $(x-3)$ is a factor of

  i $2x^3 + x^2 - 15x - 18$

  ii $x^3 - 2x^2 - 5x + 6$

b Hence, fully factorise

  i $2x^3 + x^2 - 15x - 18$

  ii $x^3 - 2x^2 - 5x + 6$

c $\dfrac{2x^3 + x^2 - 15x - 18}{x^3 - 2x^2 - 5x + 6} = \dfrac{Ax + B}{x + C}$, where A, B and C are integers.

Work out the values of A, B and C.

5 Write as a single fraction in its simplest form

a $\dfrac{x}{2} \times \dfrac{6}{5x}$ b $\dfrac{4}{x} \div \dfrac{5}{x+1}$

c $\dfrac{x+3}{4} \times \dfrac{6}{2x+6}$ d $\dfrac{x-1}{2x+4} \times \dfrac{x+2}{3x-3}$

e $\dfrac{4x+10}{x+3} \times \dfrac{3x+9}{2x+5}$ f $\dfrac{x^2-2x}{x^2-2x-15} \times \dfrac{x^2-4x-5}{x^2-x-2}$

g $\dfrac{x^2-49}{2x^2+x} \times \dfrac{2x^2-7x-4}{x^2+3x-28}$ h $\dfrac{6x^2+5x-6}{10x^2-3x-1} \div \dfrac{3x^2-8x+4}{5x^2-9x-2}$

i $\dfrac{4x^2+23x-6}{16x^2+8x-3} \div \dfrac{2x^2+17x+30}{16x^2-9}$

Chapter 2: Algebraic manipulation

6 **Reasoning**

   a   Show that $(x + 2)$ is a factor of $x^3 - 19x - 30$.

   b   Hence, fully factorise $x^3 - 19x - 30$.

   c   Hence, simplify $\dfrac{x^2 - x - 20}{x^3 - 19x - 30} \times \dfrac{2x^2 + 7x + 6}{2x^2 + 11x + 12}$.

7 **Reasoning**

   a   Show that $(2x - 1)$ is a factor of $2x^3 + x^2 - 25x + 12$.

   b   Hence, fully factorise $2x^3 + x^2 - 25x + 12$.

   c   Hence, simplify $\dfrac{3x^2 + 17x + 10}{2x^2 + 7x - 4} \div \dfrac{9x^3 - 21x^2 - 18x}{2x^3 + x^2 - 25x + 12}$.

8 Write as a single fraction in its simplest form

   a $\dfrac{x}{2} - \dfrac{x}{3}$ b $\dfrac{x}{4} + \dfrac{x}{6}$ c $\dfrac{x}{3} + \dfrac{2x}{5}$

   d $\dfrac{1}{x} + \dfrac{2}{3x}$ e $\dfrac{3}{x} + \dfrac{5}{2x}$ f $\dfrac{1}{3x} - \dfrac{2}{5x}$

   g $\dfrac{x+4}{3} + \dfrac{x+1}{2}$ h $\dfrac{x+6}{3} - \dfrac{x+5}{4}$ i $\dfrac{3x+1}{2} - \dfrac{x-7}{4}$

   j $\dfrac{1}{x+3} + \dfrac{2}{x+4}$ k $\dfrac{4}{x+1} - \dfrac{2}{x-3}$ l $\dfrac{7}{x-2} - \dfrac{5}{x-3}$

   m $2 + \dfrac{3}{x+5}$ n $1 - \dfrac{5}{x+4}$ o $\dfrac{x+1}{x-1} + \dfrac{x+2}{x+3}$

   p $\dfrac{x-1}{x+3} - \dfrac{x-2}{x+4}$ q $\dfrac{x}{x+4} - \dfrac{x+1}{x-2}$ r $1 + \dfrac{x+2}{x-5}$

Exam-style question

9 Write $\dfrac{x+5}{x-1} + \dfrac{x-5}{x+1}$ as a single fraction.

   Give your answer in its simplest form.   **(3 marks)**

10 Write as a single fraction in its simplest form

   a $\dfrac{1}{x+4} + \dfrac{1}{x^2 + 4x}$ b $\dfrac{3}{x+2} - \dfrac{1}{x^2 + 2x}$

   c $\dfrac{2}{x^2 + 3x + 2} - \dfrac{1}{x+1}$ d $\dfrac{x}{x^2 - x} + \dfrac{x+1}{x^2 - 1}$

   e $\dfrac{x^2 - x - 2}{x^2 - 4} - \dfrac{4x}{x^2 + 2x}$ f $\dfrac{x^2 - 9}{x^2 - 4x + 3} - \dfrac{x^2 + 2x}{x^2 - 3x - 10}$

   g $\dfrac{2x^2 + 3x - 2}{2x^2 + 5x - 3} + \dfrac{2x^2 + x - 3}{x^2 - 2x + 1}$ h $\dfrac{3x^2 - 10x - 8}{x^2 - 3x - 4} - \dfrac{3x^2 + 17x - 6}{x^2 + 5x - 6}$

   i $\dfrac{4x^2 - 9}{2x^2 - 7x - 15} + \dfrac{3x^2 - x - 4}{x^2 + 5x + 4}$

11 **Reasoning** Show that $\dfrac{3x^3 + 17x^2 + 10x}{2x^3 - 3x^2 + x} \div \dfrac{4x^3 + 23x^2 + 15x}{2x^3 + 7x^2 - 4x} = \dfrac{3x^2 + 14x + 8}{4x^2 - x - 3}$.

12 **Reasoning** Show that $\dfrac{2x^2 - 9x - 5}{3x^2 - 11x - 20} + \dfrac{4x^3 - 4x^2 - 3x}{4x^3 - x} = \dfrac{10x^2 - x - 13}{(2x - 1)(3x + 4)}$.

Exam-style question

13   a   Show that $x - 2$ is a factor of $x^3 - 6x^2 + 11x - 6$.   **(2 marks)**

      b   Simplify fully $\dfrac{x-2}{x^2 - 36} \div \dfrac{x^3 - 6x^2 + 11x - 6}{x+6}$.   **(4 marks)**

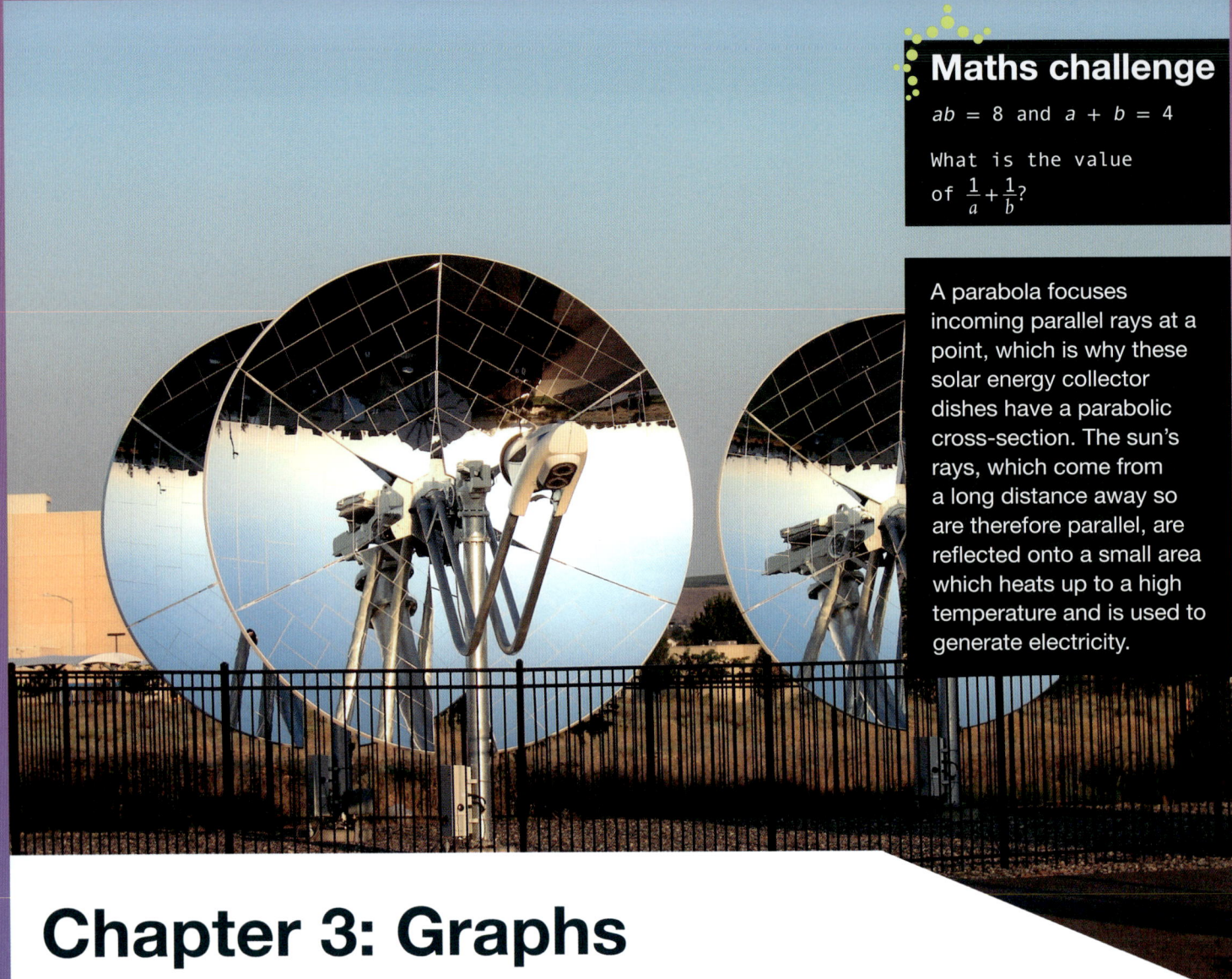

Chapter 3: Graphs

Maths challenge

$ab = 8$ and $a + b = 4$

What is the value of $\frac{1}{a} + \frac{1}{b}$?

A parabola focuses incoming parallel rays at a point, which is why these solar energy collector dishes have a parabolic cross-section. The sun's rays, which come from a long distance away so are therefore parallel, are reflected onto a small area which heats up to a high temperature and is used to generate electricity.

In this chapter you will:
- understand and use the equation of a straight line, including the forms $y - y_1 = m(x - x_1)$ and $ax + by + c = 0$
- find equations of parallel or perpendicular lines through given points
- solve problems involving parallel and perpendicular lines
- identify and interpret roots, intercepts and turning points of quadratic functions graphically and algebraically
- sketch curves defined by quadratics
- understand the links between the discriminant and no roots and real roots
- sketch graphs of cubic functions
- solve problems involving graphs of quartic functions
- sketch and interpret graphs of sine, cosine and tangent
- solve trigonometric equations

Prior knowledge
- know and use the terms 'perpendicular' and 'parallel'
- know and use the equation of a straight line $y = mx + c$
- find the gradient of a linear graph
- plot and recognise quadratic graphs
- complete the square
- know and use the factor theorem to find factors of a cubic expression
- factorise a cubic expression
- rationalise a denominator
- find sine, cosine and tangent of an angle

Chapter 3: Graphs

3.1 Linear graphs

$y = mx + c$ is the equation of a straight line, where m is the gradient, and c is the y-intercept.

$ax + by + c = 0$ is another way to write the equation of a straight line.

Example 1

a Sketch the graph of $2x + y = 6$.

b Rearrange the equation $2x + y = 6$ to the form $y = mx + c$.

c Rearrange the equation $2x + y = 6$ to the form $ax + by + c = 0$.

a When $x = 0$, $y = 6$

When $y = 0$, $2x = 6$, so $x = 3$ — Substitute $x = 0$ and $y = 0$ into the equation to find the points where the graph intercepts the axes.

Draw the axes and mark the intercepts at $x = 3$ and $y = 6$. Join with a straight line.

b $2x + y = 6$ — Subtract $2x$ from both sides to make y the subject.

$y = 6 - 2x$

$y = -2x + 6$ — Rewrite in the form $y = mx + c$. Check that this matches the sketch graph: y-intercept is 6, gradient is -2.

c $2x + y = 6$

$2x + y - 6 = 0$ — Subtract 6 from both sides. This is in the form $ax + by + c = 0$, where $a = 2$, $b = 1$ and $c = -6$.

gradient of a linear graph $= \dfrac{\text{difference in } y\text{-coordinates}}{\text{difference in } x\text{-coordinates}}$

or, using algebra, $m = \dfrac{y_2 - y_1}{x_2 - x_1}$

$y - y_1 = m(x - x_1)$ is another way to write the equation of a straight line, using the gradient and the coordinates of one point.

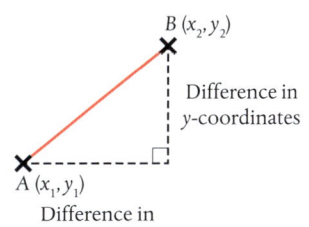

Chapter 3: Graphs

Example 2

Find the equation of the straight line through the points $(-2, 5)$ and $(4, 7)$.

Give the answer in the form $y - y_1 = m(x - x_1)$.

Find the gradient. → $m = \dfrac{y_2 - y_1}{x_2 - x_1}$

Substitute the x- and y-values into the gradient formula:
(x_1, y_1) is $(4, 7)$
(x_2, y_2) is $(-2, 5)$
→ $= \dfrac{5 - 7}{-2 - 4}$

$= \dfrac{-2}{-6}$

Divide top and bottom by -2 → $= \dfrac{1}{3}$

$y - y_1 = m(x - x_1)$

Substitute the values of m, x_1 and y_1. → $y - 5 = \dfrac{1}{3}(x - -2)$

$y - 5 = \dfrac{1}{3}(x + 2)$

Parallel lines have the same gradient, m.

When two lines are perpendicular, the product of their gradients is -1.

When a line has gradient m, a line perpendicular to it has gradient $-\dfrac{1}{m}$.

Example 3

Find the equation of the line perpendicular to $3x + 4y - 5 = 0$ that passes through $(-1, 2)$.

$3x + 4y - 5 = 0$

$4y = -3x + 5$

Rearrange into the form $y = mx + c$ → $y = -\dfrac{3}{4}x + \dfrac{5}{4}$

Find the gradient. → Gradient, $m = -\dfrac{3}{4}$

Gradient of line perpendicular to $3x + 4y - 5 = 0$ is $-\dfrac{1}{m} = \dfrac{4}{3}$

Use the gradient and the coordinates of the given point to write the equation in the form $y - y_1 = m(x - x_1)$ → $y - y_1 = m(x - x_1)$

$y - 2 = \dfrac{4}{3}(x - -1)$

$y - 2 = \dfrac{4}{3}(x + 1)$

Chapter 3: Graphs

Practice

1 Here are the equations of four straight lines.

$3x + 2y = 7$

$y = x + 2$

$y - 2 = 5(x - 1)$

$x + 3y - 7 = 0$

Which of these lines pass through (1, 2)?

Hint for Q1

Substitute the coordinates of the point ($x = 1, y = 2$) to see if they satisfy each equation.

2 Here are the graphs of lines A, B, C and D.

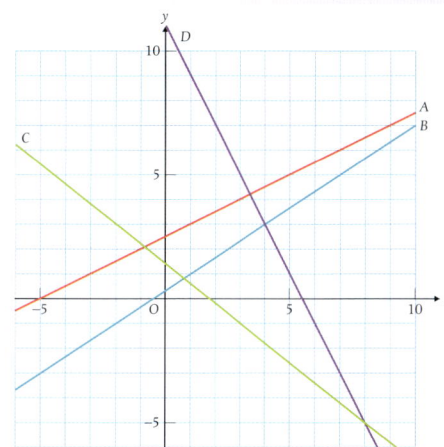

a Find the equation of line A, in the form $y - y_1 = m(x - x_1)$.

b Find the equation of line B, in the form $ax + by + c = 0$.

c Find the equation of line C, in the form $ax + by + c = 0$.

d Find the equation of line D, in the form $y - y_1 = m(x - x_1)$.

Hint for Q2

First find the gradient.

3 Here are the equations of four straight lines.

A: $4x + y - 2 = 0$

C: $y = 4x + 3$

B: $y - 3 = 4(x - 2)$

D: $y = \dfrac{x + 3}{4}$

Which of these lines are parallel?

4 Here are the equations of four straight lines.

A: $y = 3x - 7$

C: $y + 3 = 2(x + 5)$

B: $y + 4x - 7 = 0$

D: $3x + 2y - 14 = 0$

Which of these lines have the same y-intercept?

5 a Find the equation of the line parallel to $y + 3 = 2(x - 5)$ that passes through the point (4, −1).

b Find the equation of the line parallel to $y + 4x - 7 = 0$ that passes through the point (2, 6).

c Find the equation of the line parallel to $3y - 2x + 2 = 0$ that passes through the point (3, −2).

Hint for Q5a

Copy and complete the working:

Gradient = ___

$y - y_1 = m(x - x_1)$

$y - -1 = 2(x - $ ___$)$

Chapter 3: Graphs

6 a Find the gradient of a line perpendicular to the line $y + 3 = 2(x - 5)$.

   b Find the gradient of a line perpendicular to the line $y = -\frac{1}{3}x + 2$.

   c Find the gradient of a line perpendicular to the line $4x + 3y - 1 = 0$.

   d Find the gradient of a line perpendicular to the line $y - 3 = -0.4(x + 1)$.

7 a Find the equation of the line that is perpendicular to the line $y = -2x + 3$ and that passes through the point $(-2, 1)$.

   b Find the equation of the line that is perpendicular to the line $y - 2 = 4(x + 1)$ and that passes through the point $(4, 5)$.

   Give the answer in the form $ax + by + c = 0$.

   c Find the equation of the line that is perpendicular to the line $4x + 3y + 5 = 0$ and that passes through the point $(\sqrt{2}, -10)$.

Hint for Q7b
Find the equation in the form $y - y_1 = m(x - x_1)$ and rearrange into the form $ax + by + c = 0$, where a is a positive integer.

Hint for Q8
First find the gradient using $m = \frac{y_2 - y_1}{x_2 - x_1}$

8 Find the equation of the line that passes through the points $(-2, 1)$ and $(3, 11)$.

   Give the answer in the form $y - y_1 = m(x - x_1)$.

Exam-style question

9 **Problem-solving** Line L has equation $y = 3x - 7$.

   Find the equation of the line that is perpendicular to line L and passes through the point $(9, -1)$.

   Give the answer in the form $ax + by = c$. **(3 marks)**

10 Show that the points $P(3, 2)$, $Q(4, 5)$ and $R(6, 11)$ can be joined by a straight line.

11 **Problem-solving** Here are some conversion facts for temperatures:

   $32\,°F = 0\,°C$

   $212\,°F = 100\,°C$

   a Sketch a degrees Fahrenheit (°F) to degrees Celsius (°C) conversion graph. Put Fahrenheit on the horizontal axis.

   b Find the equation of the graph.

   c **Reasoning** Copy and complete this statement:
      an increase of 1°F = an increase of ___ °C

   d **Reasoning** Find the y-intercept and explain what it represents in the context of the graph.

12 **Reasoning** The points $A(-4, 0)$, $B(0, 3)$ and $C(-2, -1)$ are three vertices of a triangle.

   Show that triangle ABC is a right-angled triangle.

 Talking point
In Q10, there are two possible equations for the line, depending on which pair of points you use. Does it matter which equation you find?

Hint for Q11d
The y-intercept is the temperature in °C equivalent to ___°F

Chapter 3: Graphs

3.2 Quadratic graphs

A quadratic graph is a parabola (U-shaped curve).

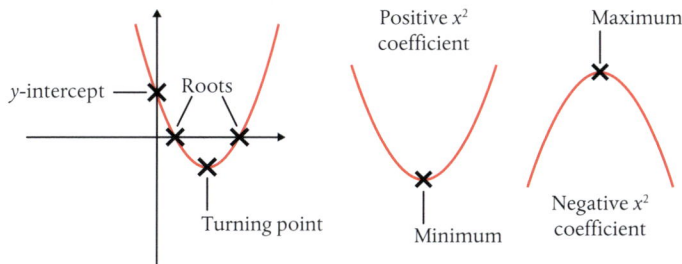

The roots of an equation, where $y = 0$, are its solutions.

Example 1

Here is the graph of $y = x^2 + 3x - 5$.

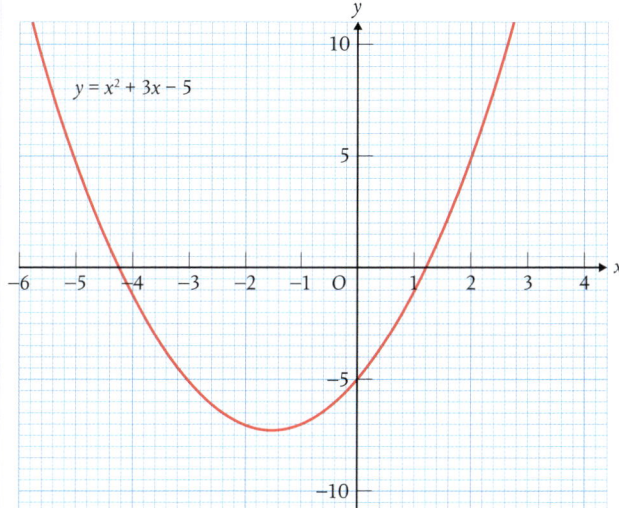

By drawing a suitable line, use this graph to estimate the solutions to $y = x^2 + x - 8$.

Solutions are where:

$x^2 + x - 8 = 0$ ———— Rearrange so that the LHS is equal to $x^2 + 3x - 5$

$x^2 + 3x - 8 = 2x$ ———— Add $2x$ to both sides.

$x^2 + 3x - 5 = 2x + 3$ ———— Add 3 to both sides.

Chapter 3: Graphs

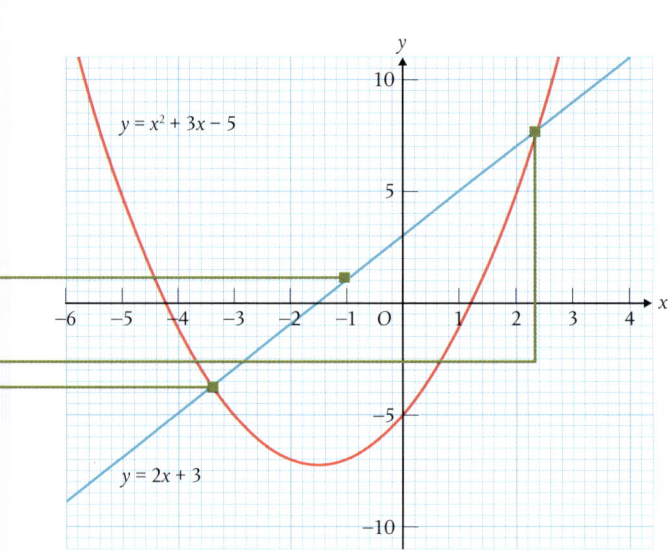

Draw the line $y = 2x + 3$ on the graph.

Estimate the x-values where the line and curve intersect.

$x = -3.4$ (1 d.p.) and $x = 2.4$ (1 d.p.)

One way to find the roots of a quadratic equation is to use the quadratic formula.

$$x = \frac{-b \pm \sqrt{b^2 - 4ac}}{2a}$$

Example 2

Find the roots of $3x^2 + 2x - 6$ to one decimal place.

$$x = \frac{-b \pm \sqrt{b^2 - 4ac}}{2a}$$

$$x = \frac{-2 \pm \sqrt{2^2 - 4 \times 3 \times -6}}{2 \times 3}$$

$$x = \frac{-2 \pm \sqrt{4 + 72}}{6}$$

$$x = \frac{-2 \pm \sqrt{76}}{6}$$

$x = 1.1$ or $x = -1.8$ (1 d.p.)

Substitute $a = 3$, $b = 2$ and $c = -6$ into the quadratic formula.
Use a calculator and round to 1 decimal place.

The discriminant is the expression inside the square root, in the quadratic formula.

$$x = \frac{-b \pm \sqrt{b^2 - 4ac}}{2a}$$ ⎯⎯⎯⎯ discriminant

- When $b^2 - 4ac > 0$, there are two real roots (as in Example 2).

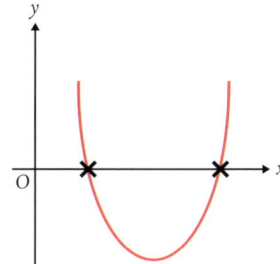

- When $b^2 - 4ac = 0$, there is one real root. For example, for the equation $x^2 - 2x + 1$:

$$x = \frac{-2 \pm \sqrt{2^2 - 4 \times 1 \times 1}}{2 \times 1} = \frac{-2 + \sqrt{0}}{2} = -1$$

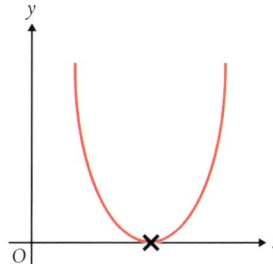

- When $b^2 - 4ac < 0$ there are no real roots, because there is no true square root of a negative number.

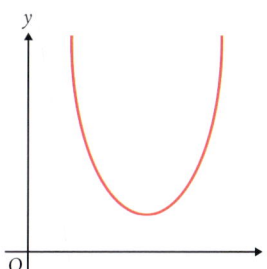

Example 3

Show that the equation $9x^2 + 6x + 1 = 0$ has exactly one real root.

$b^2 - 4ac = 6^2 - 4 \times 9 \times 1$ ⎯⎯⎯⎯ Calculate the discriminant.

$= 36 - 36$

$= 0$

The discriminant $= 0$, so the equation has exactly one real root.

Chapter 3: Graphs

When a quadratic equation is written in completed square form $y = a(x + h)^2 + k$, the turning point is $(-h, k)$.

Example 4

Find the turning point of the quadratic function $3x^2 + 6x + 5$, and state whether it is a maximum or minimum.

$3x^2 + 6x + 5 = 3(x^2 + 2x) + 5$

$ = 3((x+1)^2 - 1) + 5$

$ = 3(x+1)^2 - 3 + 5$

$ = 3(x+1)^2 + 2$

Turning point $= (-1, 2)$ It is a minimum.

- Write in the form $a\left(x^2 + \frac{b}{a}x\right) + c$, then complete the square for $x^2 + \frac{b}{a}x$

- This is in the form $a(x + h)^2 + k$, so the turning point is $(-h, k)$

- Quadratic graphs with positive a are U-shaped so the turning point is a minimum.

Practice

1 Here is a graph of $y = x^2 + 4x - 1$.

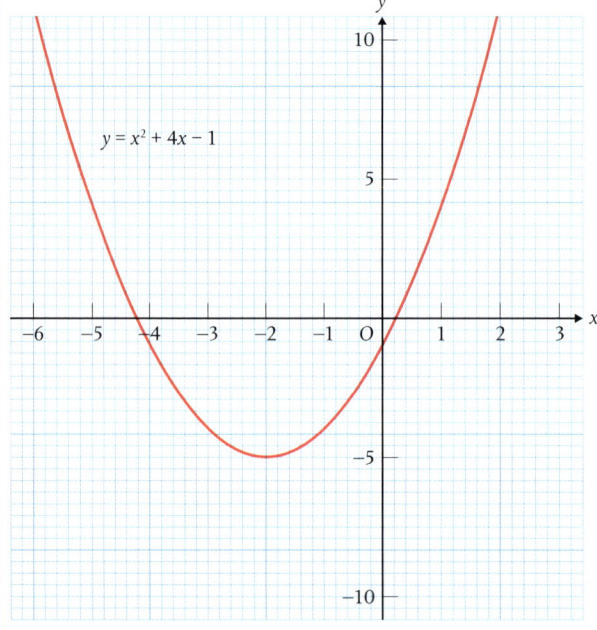

By drawing a suitable line, use this graph to estimate the solutions to $y = x^2 + 4x - 4$.

2. Here is a graph of $y = x^2 - 2x + 3$.

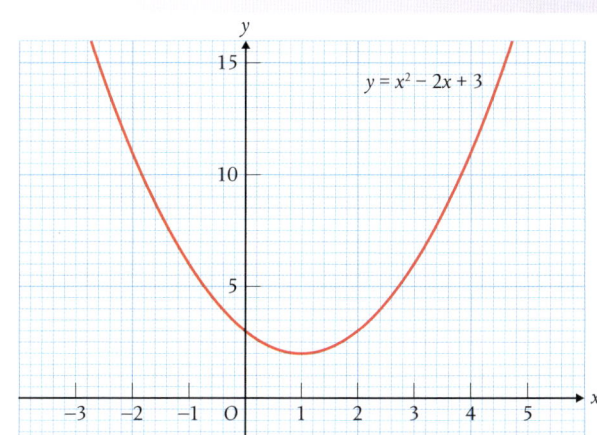

By drawing a suitable line, use this graph to estimate the solutions to $y = x^2 - 4x + 2$.

3. Find the roots of each quadratic equation, to two decimal places.

   a $y = 2x^2 - 10x + 5$
   b $y = 3x^2 - 6x - 5$
   c $y = 5x^2 + 3x - 1$

   Hint for Q3
   Use the quadratic formula.

4. Find the roots of each quadratic equation. Give the answers in surd form.

   a $y = 2x^2 + 7x + 1$
   b $y = 2x^2 + 8x + 5$
   c $y = 3x^2 + 2x - 2$

5. Find the number of real roots for each equation.

   a $y = 5x^2 + 4x - 3$
   b $y = 6x^2 + 2x + 5$
   c $y = 3x^2 + 7x + 4$
   d $y = 4x^2 - 3$
   e $y = 4x^2 + 9x + 7$
   f $y = 9x^2 + 12x + 4$

6. Find the turning point of each quadratic function, and state whether it is a maximum or minimum.

   a $y = 5x^2 + 20x + 21$
   b $y = 2x^2 + 3x - 1$
   c $y = -3x^2 + 12x - 9$
   d $y = -3x^2 + 6x + 7$

   Talking point
   Explain how to use the symmetry of a quadratic curve to find the x-coordinate of the turning point, given the roots.
   Explain how to find the y-coordinate of the turning point.

7. a Find the roots of the equation $y = 2x^2 + 4x - 5$. Give the answers to 1 decimal place.

   b Find the turning point of the graph of $y = 2x^2 + 4x - 5$ and state whether it is a maximum or a minimum.

   c Find the y-intercept of the graph of $y = 2x^2 + 4x - 5$.

   d Using the results from parts **a–c**, sketch the graph of $y = 2x^2 + 4x - 5$.

   Hint for Q7c
   At the y-intercept, $x = 0$.

   Hint for Q7d
   In a sketch graph, label the coordinates of the y-intercept, the roots and the turning point.

8. Sketch the graph of $y = 3x^2 - 12x + 7$.

   Hint for Q8
   Follow the steps in Q7.

9. Sketch the graph of $y = -2x^2 + 4x + 1$, labelling the roots in surd form.

Chapter 3: Graphs

Exam-style question

10 Sketch the graph of

$y = 2x^2 - 8x - 5$

Show the coordinates of the turning point and the exact coordinates of any intercepts with the coordinate axes. **(5 marks)**

11 Here is a quadratic graph.

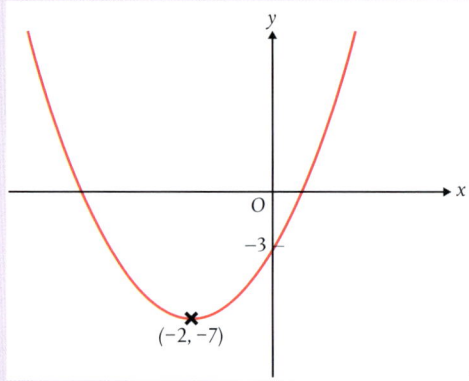

a Find the equation of the graph.

b Find the intercepts with the x-axis.

12 **Reasoning** Laila draws this sketch of the graph of $y = 2x^2 - 5x + 6$.

The sketch is not correct.

Hint for Q12

Use the discriminant.

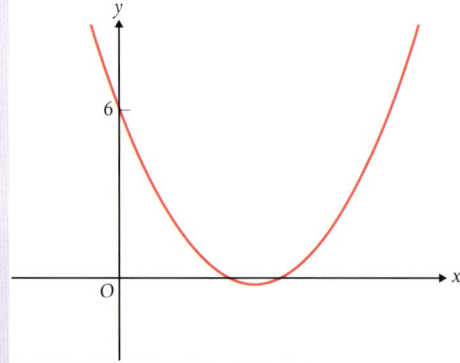

Explain why this sketch graph cannot be the graph of $y = 2x^2 - 5x + 6$.

13 Reasoning Here are four quadratic graphs.

A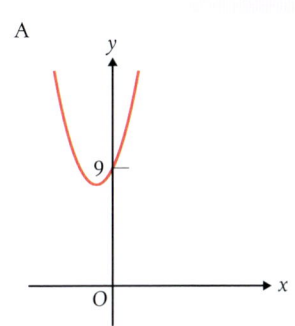

B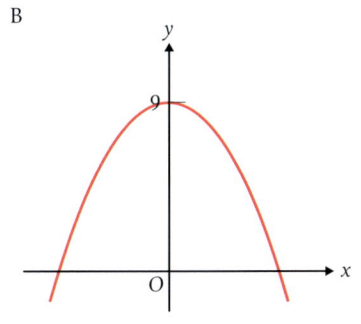

C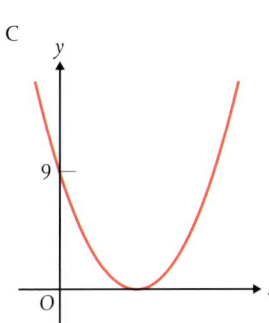

D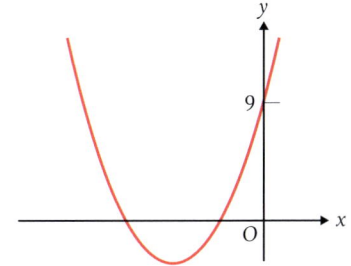

Here are the equations of the four graphs.

$y = -x^2 + 9$

$y = x^2 - 6x + 9$

$y = x^2 + 7x + 9$

$y = 3x^2 + 4x + 9$

By considering the number of roots, or otherwise, match each graph to its equation.

14 The quadratic equation $2x^2 - 5x + p = 0$ has exactly one root.

Work out the value of p.

Chapter 3: Graphs

3.3 Cubic and quartic graphs

A cubic function has one, two or three roots. The roots are the x-values when $y = 0$.

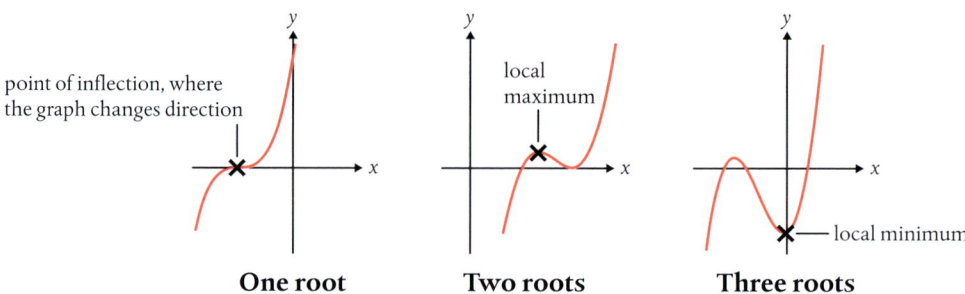

point of inflection, where the graph changes direction

local maximum

local minimum

One root   **Two roots**   **Three roots**

The graph of $y = -x^3$ is a reflection of the graph of $y = x^3$ in the y-axis.

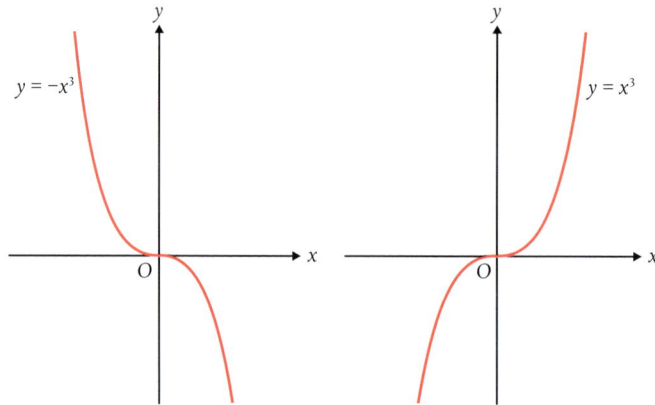

$y = -x^3$ $y = x^3$

The y-intercept of a graph is the y-value when $x = 0$.

Chapter 3: Graphs

Example 1

Sketch the graph of

a $y = x(x-3)(x+2)$ **b** $y = (x+1)^2(x-4)$

a $y = x(x-3)(x+2)$ — Substitute $x = 0$ into the equation to find the y-intercept.

When $x = 0$, $y = 0$.

So, y-intercept $= 0$

$x(x-3)(x+2) = 0$ — Find the roots. The roots are the x-values when $y = 0$.

$x = 0$ or $x = 3$ or $x = -2$

$x(x-3)(x+2) = x(x^2 - x - 6)$ — Expand the expression to find the coefficient of x^3.

$\qquad = x^3 - x^2 - 6x$

The coefficient of x^3 is positive, so the graph has the shape 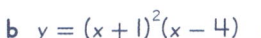.

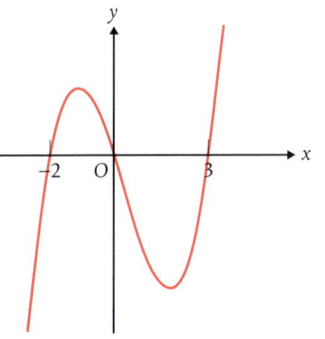

Mark the roots and y-intercept on the axes, and sketch the graph.

b $y = (x+1)^2(x-4)$ — Substitute $x = 0$ into the equation to find the y-intercept.

When $x = 0$, $y = 1 \times -4 = -4$

So, y-intercept $= -4$

$(x+1)^2(x-4) = 0$ — Find the roots. The roots are the x-values when $y = 0$.

$x = -1$ or $x = 4$

$(x^2 + 2x + 1)(x-4) = x^3 - 4x^2 + 2x^2 - 8x + x - 4$ — Expand the expression to find the coefficient of x^3.

$\qquad = x^3 - 2x^2 - 7x - 4$

The coefficient of x^3 is positive, so the graph has the shape .

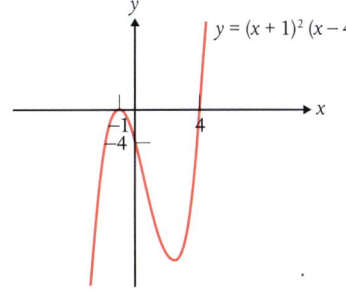

Mark the roots and y-intercept on the axes, and sketch the graph.

💬 **Talking point**

Is it necessary to expand the whole expression to find the y-intercept?

Chapter 3: Graphs

Example 2

Here is the graph of $y = x^3 - 2x - 5$.

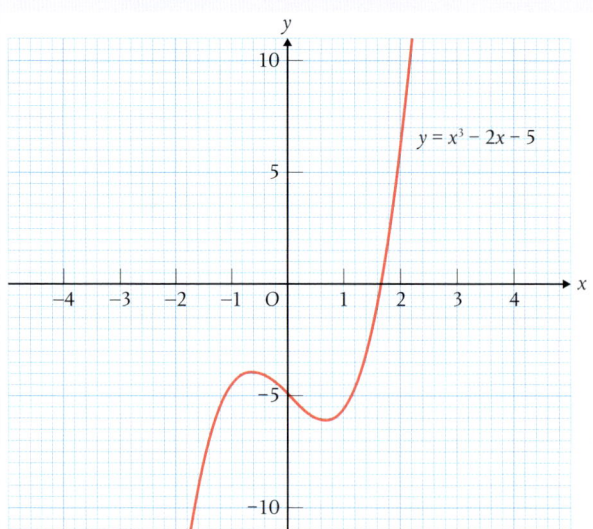

By drawing a suitable line, use this graph to estimate the solutions to $y = x^3 - 6x - 4$.

Solutions are where:

- Rearrange so that the LHS is equal to $x^3 - 2x - 5$. → $x^3 - 6x - 4 = 0$
- Add $4x$ to both sides. → $x^3 - 2x - 4 = 4x$
- Subtract 1 from both sides. → $x^3 - 2x - 5 = 4x - 1$

Draw the line $y = 4x - 1$ on the graph.

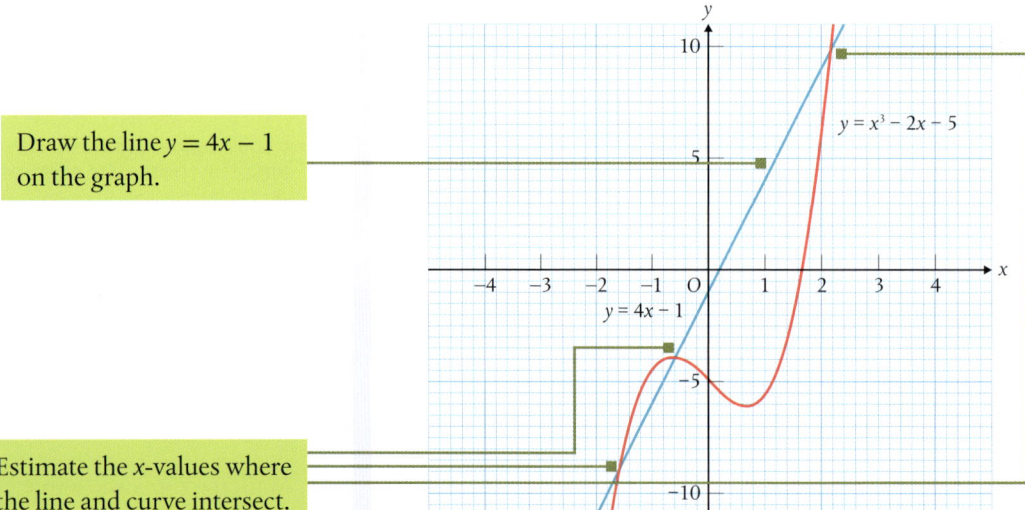

Estimate the x-values where the line and curve intersect.

Estimates for the solutions to the equation $x^3 - 6x - 4 = 0$ are -1.6 (1 d.p.), -0.7 (1 d.p.), 2.1 (1 d.p.)

Chapter 3: Graphs

A quartic expression has an x^4 term and no higher power of x. For example, $x^4 + 2x^3 - 5x + 1$.

A quartic graph with two repeated roots has a W-shape.

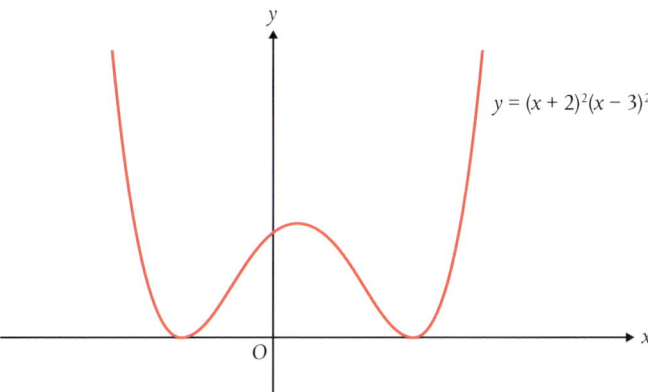

$y = (x + 2)^2(x - 3)^2$

Example 3

a Plot the graph of $-x^4 - 4x^3 - x^2 + 6x$ for $-4 \leq x \leq 2$ to show the intercepts with the axes.

b Using the graph, write the quartic expression $-x^4 - 4x^3 - x^2 + 6x$ in the form $-x(x + a)(x + b)(x + c)$ where a, b and c are integers.

a

x	−4	−3	−2	−1	0	1	2
y	−40	0	0	−4	0	0	−40

Make a table of values and plot the graph, or use a graph plotting program.

Show the intercepts with the x- and y-axes accurately. For this question it is not necessary to work out the coordinates of the two maxima or the minimum point.

b Roots are $x = -3$, $x = -2$, $x = 0$, $x = 1$

$-x^4 - 4x^3 - x^2 + 6x = -x(x + 3)(x + 2)(x - 1)$

The roots of the equation are the x-values where $y = 0$.

Practice

1. Sketch the graph of the equation $y = (x + 2)(x - 3)(x - 5)$.

2. Sketch the graph of the equation $y = x(x - 4)^2$.

3. Sketch the graph of the equation $y = -(x - 1)(x + 3)^2$.

4. Here is the graph of $y = x^3 - 3x + 1$.

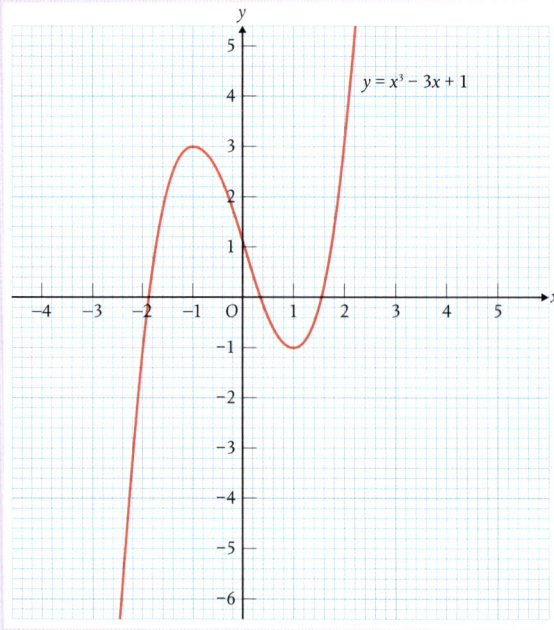

By drawing a suitable line, use this graph to estimate the solution to $x^3 - 4x + 4 = 0$.

5. Here is the graph of $y = -x^3 + 2x + 3$.

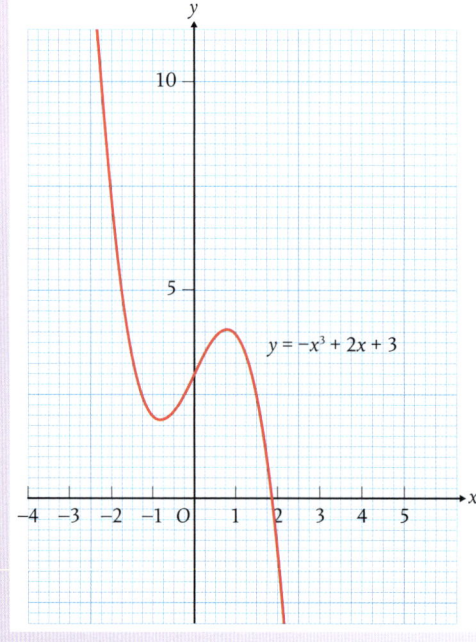

By drawing a suitable line, use this graph to estimate the solution to $-x^3 + 4x - 1 = 0$.

6   a   Use the fourth row of Pascal's triangle, 1, 3, 3, 1, to expand this cubic expression $(x + 5)^3$.

    b   Sketch the graph of $y = (x + 5)^3$.

Hint for Q6b

What shape is a cubic graph with exactly one root?

7   **Problem-solving** Find the equation of each cubic graph. Give the answers in the form $y = ax^3 + bx^2 + cx + d$.

a

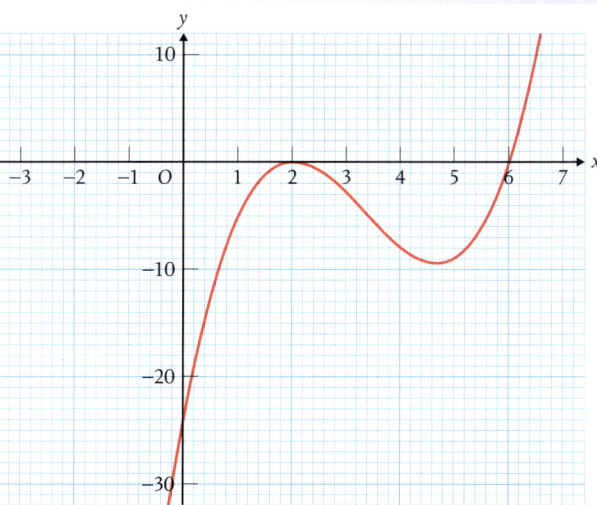

b

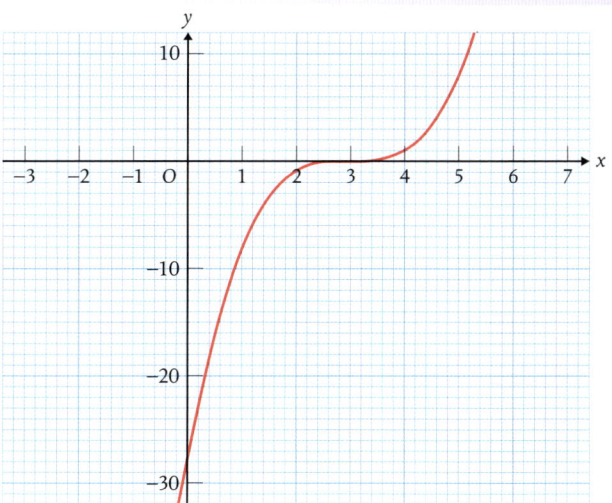

8   Sketch the graph of the equation $y = (2 - x)(x + 3)(2x - 1)$.

9 a Use the factor theorem to show that $(x-1)$ is a factor of $x^3 + 6x^2 + 3x - 10$.

   b Factorise $x^3 + 6x^2 + 3x - 10$ into the form $(x-1)(x^2 + bx + c)$.

   c Hence write $x^3 + 6x^2 + 3x - 10$ in the form $(x-1)(x+m)(x+n)$.

   d Sketch the graph of the cubic function $y = x^3 + 6x^2 + 3x - 10$.

10 **Problem-solving** The graph of $y = (4x-1)(x+3)(x+1)$ has turning points at $(-2.2, 9.4)$ and $(-0.3, -4.2)$.

   Sketch the graph of $y = (4x-1)(x+3)(x+1)$ showing the coordinates of the turning points and the coordinates of any intercepts with the coordinate axes.

11 **Problem-solving** Here is a sketch graph of $y = x^3 - 7x - 6$.

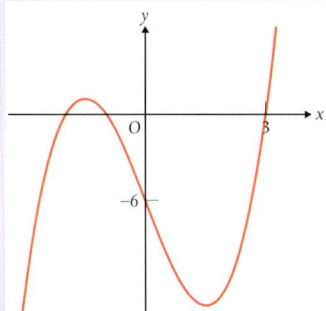

   Find the coordinates of the two unlabelled x-intercepts.

Hint for Q12a
Use a graph plotting program or a table of values.

12 a Plot the graph of $y = x^4 + 2x^3 - 13x^2 - 14x + 24$ for $-4 \le x \le 4$ to show the x- and y-intercepts.

   b Using the graph, write the quartic expression $y = x^4 + 2x^3 - 13x^2 - 14x + 24$ in the form $(x+a)(x+b)(x+c)(x+d)$, where a, b, c and d are integers.

13 a Plot the graph of $y = x^4 - 2x^3 - 3x^2 + 4x + 4$ for $-3 \le x \le 3$ to show the x- and y-intercepts.

   b Using the graph, write the quartic expression $y = x^4 - 2x^3 - 3x^2 + 4x + 4$ in the form $(x+a)^2(x+b)^2$, where a and b are integers.

14 **Reasoning** Sketch the graph of $(x-1)^2(x-3)^2$.
   Label the x- and y-intercepts.
   Label the coordinates of the local maximum point.

Exam-style questions

15 Sketch the graph of
$y = (2x - 5)(x + 2)(3 - x)$
showing the exact coordinates of any intercepts with the coordinate axes.

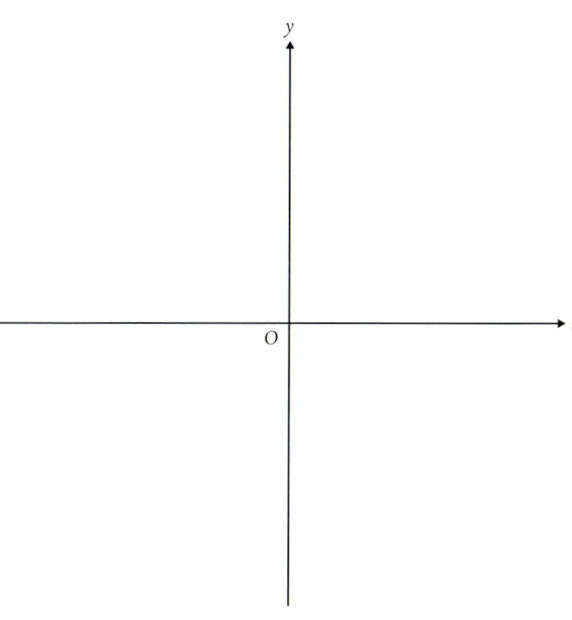

(4 marks)

16 Sketch the graph of
$y = (x + 5)(x - 3)(x + 2)^2$
showing the exact coordinates of any intercepts with the coordinate axes.

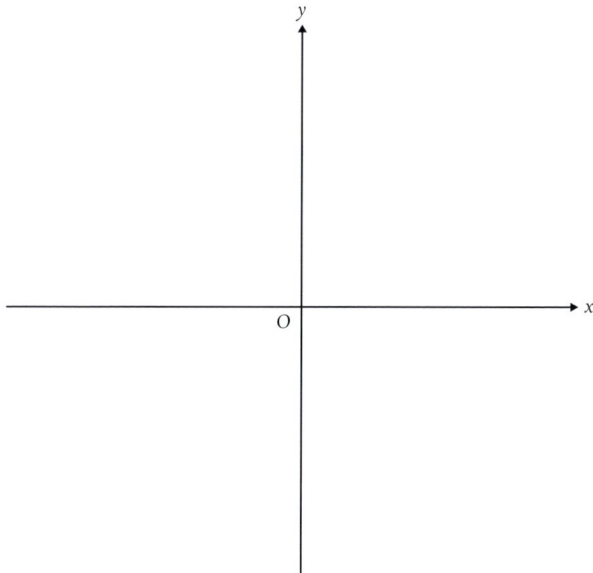

(4 marks)

Chapter 3: Graphs

3.4 Trigonometric graphs

It is important to learn these values of the trigonometric functions sine, cosine and tangent:

$\sin 0° = 0$ $\qquad\qquad \cos 0° = 1 \qquad\qquad \tan 0° = 0$

$\sin 90° = 1 \qquad\qquad \cos 90° = 0 \qquad\qquad \tan 90°$ is undefined

- Use an isosceles right-angled triangle with perpendicular sides of 1 cm to find the exact values of sin 45°, cos 45° and tan 45°.

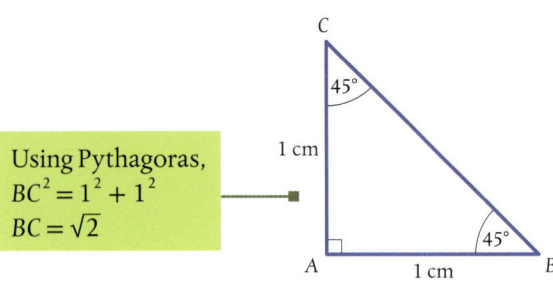

Using Pythagoras,
$BC^2 = 1^2 + 1^2$
$BC = \sqrt{2}$

$\sin 45° = \dfrac{1}{\sqrt{2}} \qquad\qquad \cos 45° = \dfrac{1}{\sqrt{2}} \qquad\qquad \tan 45° = 1$

- Use an equilateral triangle with sides of 2 cm to find the exact values of sin 60°, cos 60°, tan 60°, sin 30°, cos 30° and tan 30°.

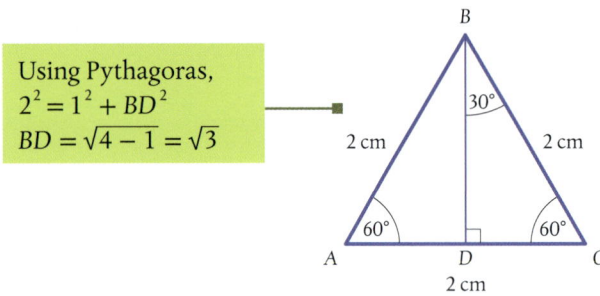

Using Pythagoras,
$2^2 = 1^2 + BD^2$
$BD = \sqrt{4-1} = \sqrt{3}$

$\sin 60° = \dfrac{\sqrt{3}}{2} \qquad\qquad \cos 60° = \dfrac{1}{2} \qquad\qquad \tan 60° = \sqrt{3}$

$\sin 30° = \dfrac{1}{2} \qquad\qquad \cos 30° = \dfrac{\sqrt{3}}{2} \qquad\qquad \tan 30° = \dfrac{1}{\sqrt{3}}$

$\sin 60° = \dfrac{\sqrt{3}}{2} = \cos 30°$

$\sin 30° = \dfrac{1}{2} = \cos 60°$

Generalising,

$\sin x° = \cos(90 - x)°$

Here are the sine, cosine and tangent graphs.

The **period** of a graph is the distance between two repeating points on the graph function.

$y = \sin x°$ $\qquad\qquad\qquad\qquad y = \cos x° \qquad\qquad\qquad\qquad y = \tan x°$

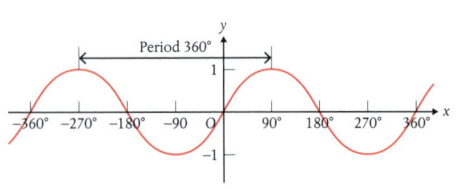 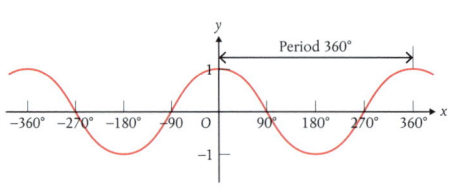 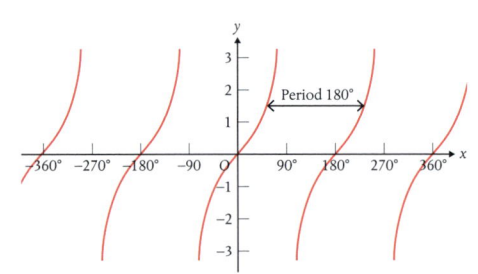

Chapter 3: Graphs

Example 1

Here is a graph of sin x.

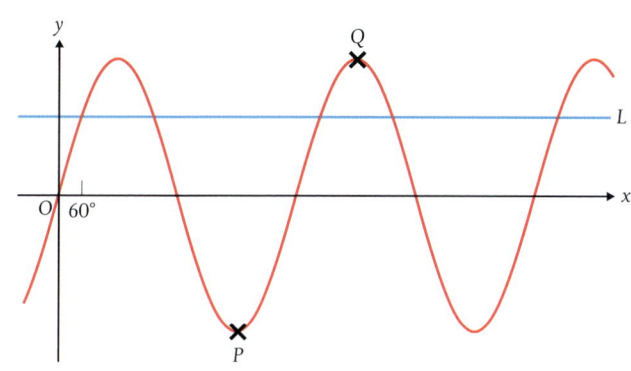

a Find the coordinates of the points marked

   i P   ii Q

b Find the equation of line L.

c Write down three angles between 90° and 540° that satisfy the equation $\sin x° = \frac{\sqrt{3}}{2}$.

a i y-coordinate of P is –1

> The graph of $y = \sin x$ has maximum value 1 and minimum value –1.

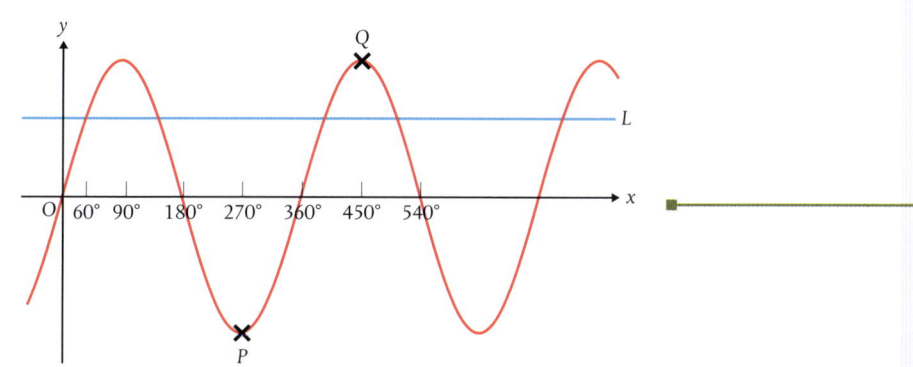

> $\sin 90° = 1$
> Use the symmetry of the graph to label the x-axis.

P = (270°, –1)

ii Q = (450°, 1)

b Line L intercepts the graph where x = 60°.

$\sin 60° = \frac{\sqrt{3}}{2}$

Equation of line L is $y = \frac{\sqrt{3}}{2}$

> L passes through the point $\left(60°, \frac{\sqrt{3}}{2}\right)$

3.4 Trigonometric graphs

49

Chapter 3: Graphs

c All the points where line L intercepts the graph have sine value $\frac{\sqrt{3}}{2}$

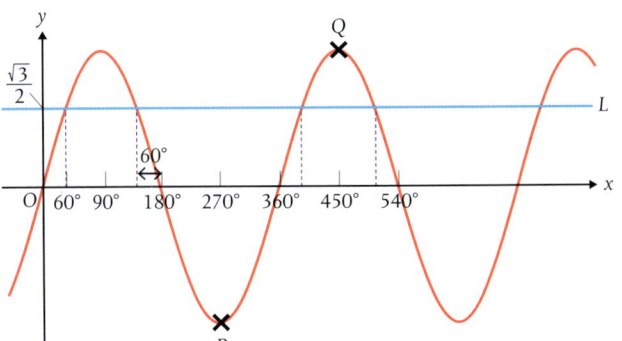

> Use the symmetry of the graph to find the angles x that satisfy $\sin x = \frac{\sqrt{3}}{2}$.

Three angles between 90° and 540° that satisfy the equation $\sin x° = \frac{\sqrt{3}}{2}$ are 120°, 360° + 60° = 420°, 540° − 60° = 480°

Example 2

a Solve $\tan x° = \frac{1}{\sqrt{3}}$ for all values of x such that $-360° \leq x \leq 360°$.

b Solve $\tan x° = -\frac{1}{\sqrt{3}}$ for all values of x such that $-360° \leq x \leq 360°$.

a $\tan x° = \frac{1}{\sqrt{3}}$, so $x = 30°$

> Look at the exact values for tan, above.

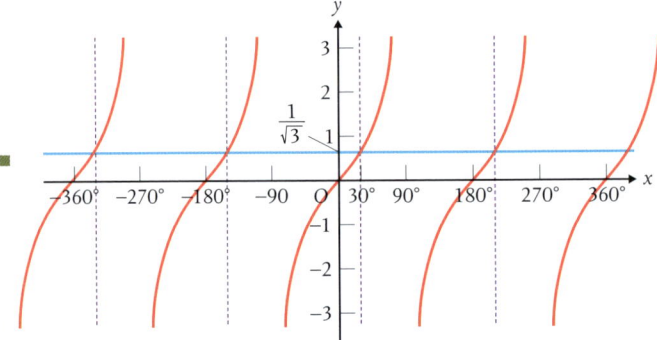

> Draw the line $\tan x° = \frac{1}{\sqrt{3}}$ on the graph of tan x, to find other values of x that satisfy the equation.

$\tan x° = \frac{1}{\sqrt{3}}$ has solutions $x = -330°, -150°, 30°, 210°$ for $-360° \leq x \leq 360°$.

b $\tan x° = -\frac{1}{\sqrt{3}}$, so $x = -30°$

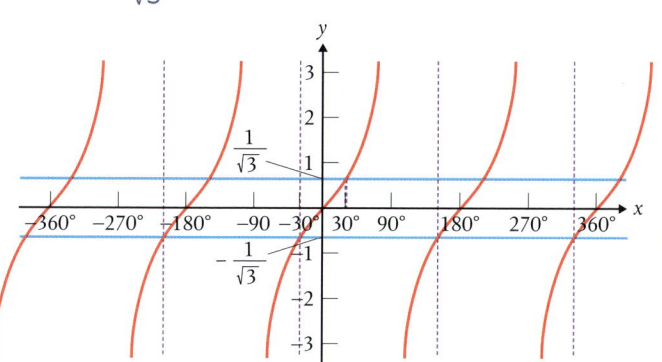

Use the symmetry of the graph to draw the line $\tan x° = -\frac{1}{\sqrt{3}}$ on the graph of $\tan x$, to find other values of x that satisfy the equation.

$\tan x° = -\frac{1}{\sqrt{3}}$ has solutions $x = -210°, -30°, 150°, 330°$ for $-360° \leq x \leq 360°$.

Practice

1 Sketch the graph of $y = \sin x°$ for $-360° \leq x \leq 720°$.

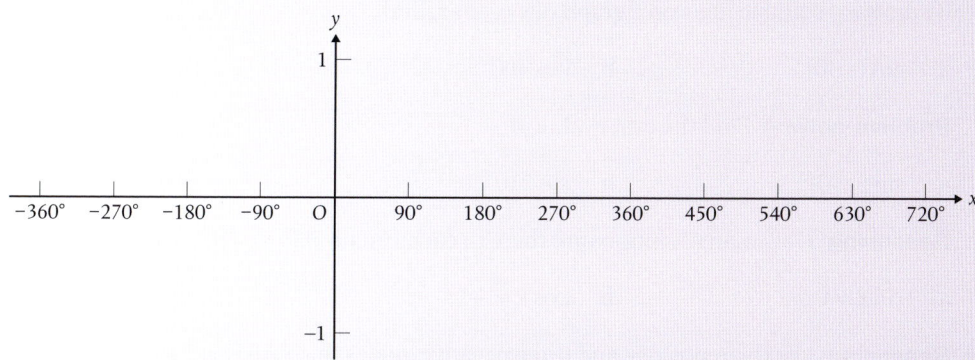

   Label the maximum and minimum values on the y-axis.

2 Sketch the graph of $y = \cos x°$ for $-360° \leq x \leq 720°$.

   Use a copy of the axes in Q1.

   Label the maximum and minimum values on the y-axis.

3 Rationalise the denominators of these exact values:

   a $\tan 30° = \frac{1}{\sqrt{3}}$ **b** $\sin 45° = \frac{1}{\sqrt{2}}$

   Hint for Q3

   To rationalise the denominator of $\frac{a}{\sqrt{b}}$, multiply by $\frac{\sqrt{b}}{\sqrt{b}}$.

Chapter 3: Graphs

4 Here is a graph of $\cos x$.

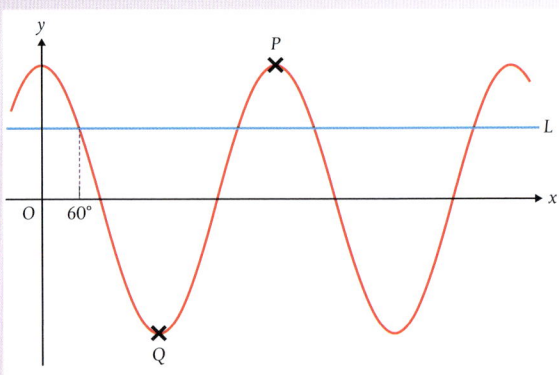

   a Find the coordinates of the points marked

      i P   ii Q

   b Find the equation of line L.

   c Write down three angles between 0° and 540° that satisfy the equation $\cos x° = \frac{1}{2}$.

Hint for 5

Use a sketch graph of $\tan x$.

5 Solve $\tan x° = 1$ for all values of x such that $-360° \leq x \leq 360°$.

6 Solve $\sin x° = \frac{\sqrt{2}}{2}$ for all values of x such that $-360° \leq x \leq 360°$.

💬 **Talking point**

Explain the relationship between the answers to **Q5** and **Q6**.

7 By drawing suitable lines on a graph of $y = \cos x$, find

   a $\cos(-30)°$ b $\cos 300°$

Hint for 7a

Use the symmetry of the graph to find another angle y such that $\cos y = \cos(-30)°$ is an exact value.

8 **Problem-solving** Find the exact value of

   a $\sin(-30)°$ b $\sin 210°$

9 **Reasoning** How many solutions are there to each equation for $-720° \leq x \leq 360°$?

   a $\sin x = 0$ b $\cos x = -1$ c $\tan x = 5$

10 **Reasoning** List three positive and three negative angles that satisfy $\cos x = 0$.

11 **Reasoning** $\sin 56° = 0.829$ (3 d.p.)

List four positive and two negative angles x, $-360° \leq x \leq 540°$, that satisfy $\sin x = 0.829$ (3 d.p.).

12 Solve $2\cos x° = \sqrt{3}$ for all values of x such that $-360° \leq x \leq 360°$.

13 $\sin x° = -\frac{\sqrt{3}}{2}$ has two possible solutions such that $0° \leq x \leq 360°$.

Find the two possible values of $\cos x$.

Exam-style question

14 Given that $\sin 30° = \frac{1}{2}$

Solve $\sin x = \frac{1}{2}$ for all values of x such that $-360° \leq x \leq 360°$.   **(2 marks)**

15 Solve $\cos x = \frac{\sqrt{2}}{2}$ for all values of x such that $-360° \leq x \leq 360°$ **(2 marks)**

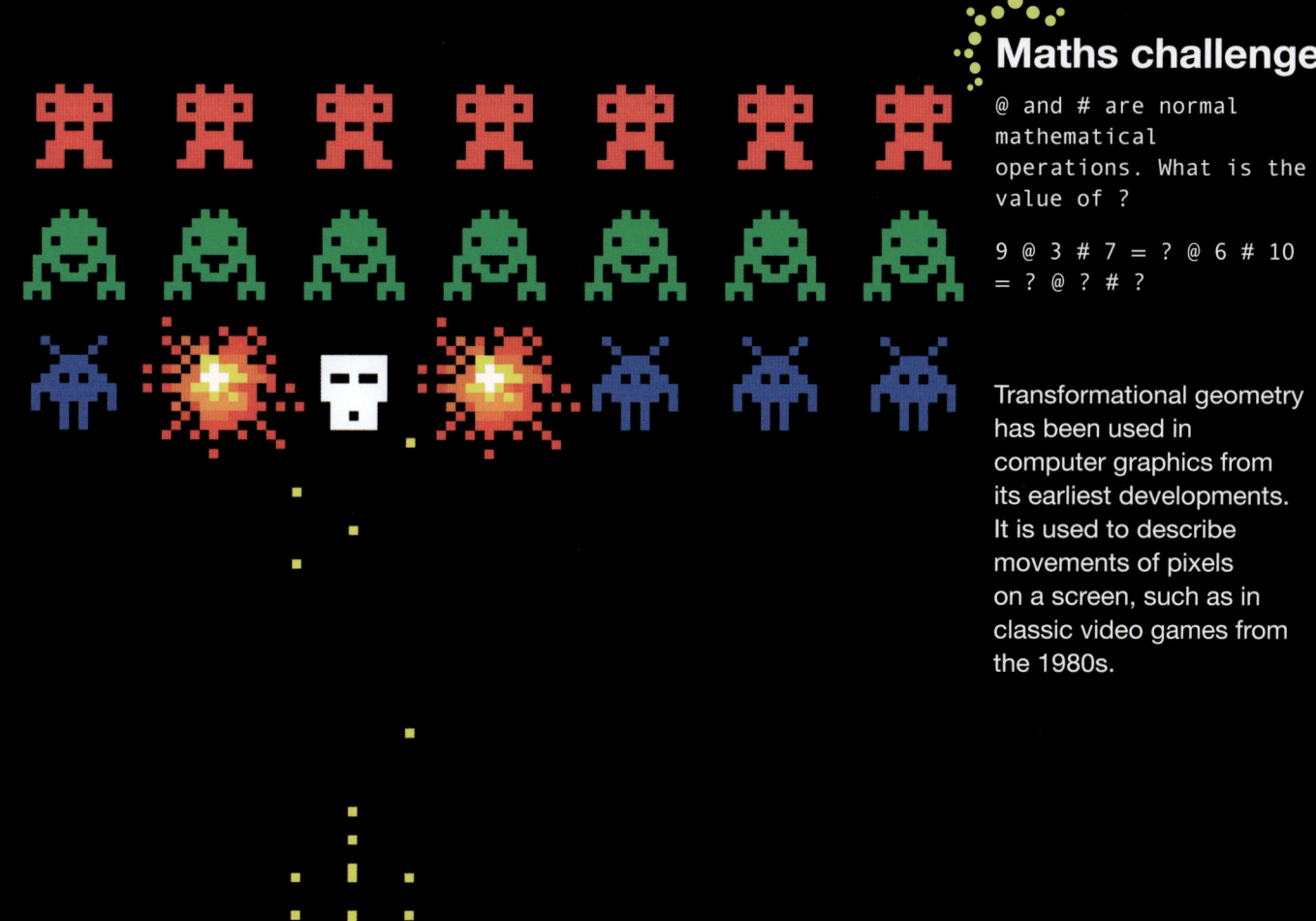

Maths challenge

@ and # are normal mathematical operations. What is the value of ?

9 @ 3 # 7 = ? @ 6 # 10 = ? @ ? # ?

Transformational geometry has been used in computer graphics from its earliest developments. It is used to describe movements of pixels on a screen, such as in classic video games from the 1980s.

Chapter 4: More graphs

In this chapter you will:
- translate, reflect and stretch graphs
- understand and use the coordinate geometry of a circle
- find the equation of a radius and tangent to a circle
- sketch, recognise and use the graphs of exponential and reciprocal functions
- estimate the gradient at a point on a curve
- estimate the area under a curve

Prior knowledge
- sketch and recognise quadratic, cubic and trigonometric graphs
- use and find the equation of a circle with centre (0, 0) and radius r
- complete the square
- use the equations of straight-line graphs
- identify and draw tangents to curves
- find the area of a trapezium

Chapter 4: More graphs

4.1 Translating and reflecting graphs

The graph of $y = f(x) + a$ is a translation of the graph $y = f(x)$ by the vector $\begin{pmatrix} 0 \\ a \end{pmatrix}$.

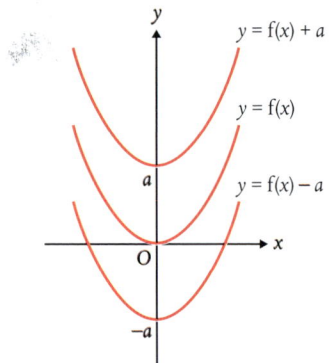

The graph of $y = f(x + a)$ is a translation of the graph $y = f(x)$ by the vector $\begin{pmatrix} -a \\ 0 \end{pmatrix}$.

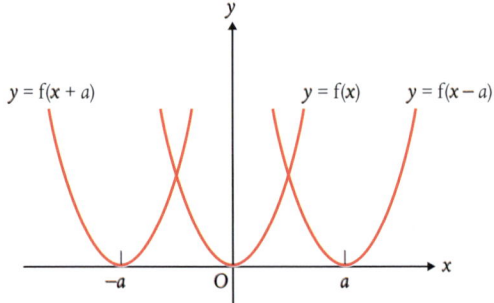

The graph of $y = -f(x)$ is a reflection of the graph of $y = f(x)$ in the x-axis.

The graph of $y = f(-x)$ is a reflection of the graph of $y = f(x)$ in the y-axis.

Chapter 4: More graphs

Example 1

$f(x) = x^3$

This is the graph of $y = f(x)$.

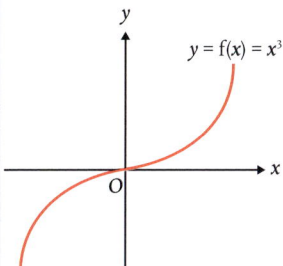

Sketch the graphs of

a $y = f(x) - 3$

b $y = f(x + 2)$

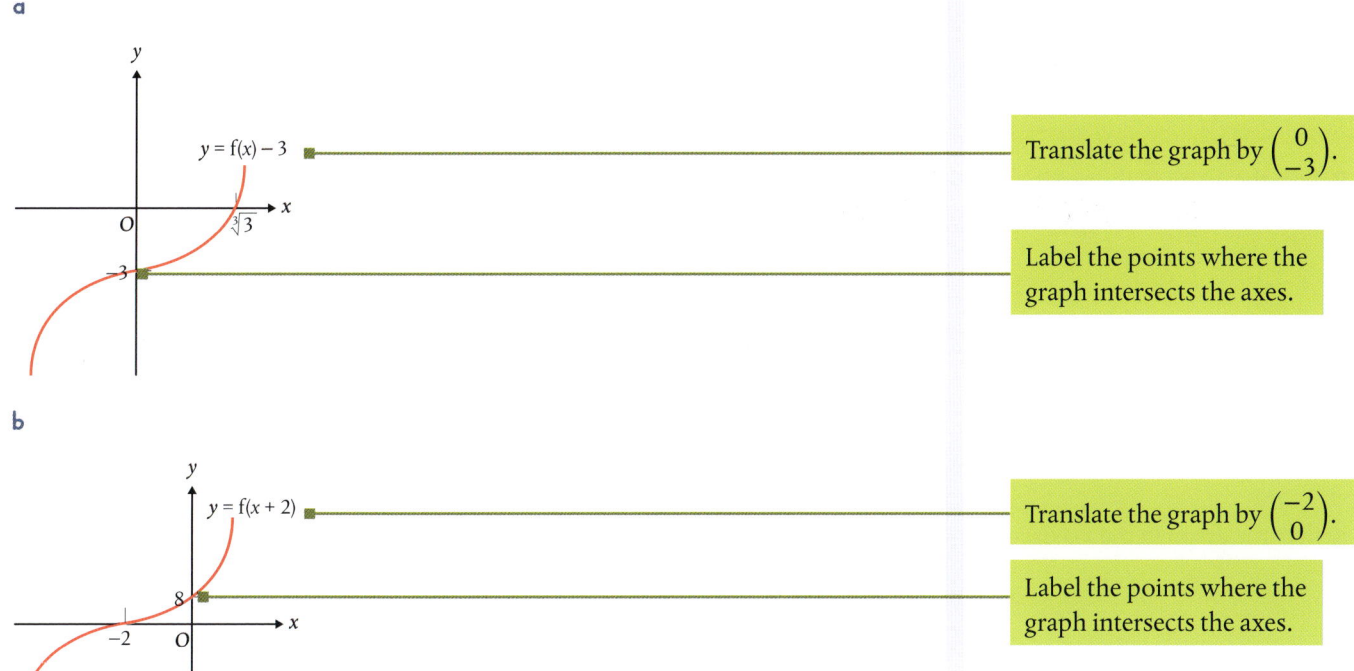

a Translate the graph by $\begin{pmatrix} 0 \\ -3 \end{pmatrix}$.

Label the points where the graph intersects the axes.

b Translate the graph by $\begin{pmatrix} -2 \\ 0 \end{pmatrix}$.

Label the points where the graph intersects the axes.

4.1 Translating and reflecting graphs

Chapter 4: More graphs

Example 2

$f(x) = x(x + 4)$

Sketch each graph.

a $y = f(x)$ **b** $y = f(-x)$ **c** $y = -f(x)$

Expand the brackets and then complete the square to find the coordinates of the turning point.

a $x(x + 4) = x^2 + 4x = (x + 2)^2 - 4$

Turning point at $(-2, -4)$

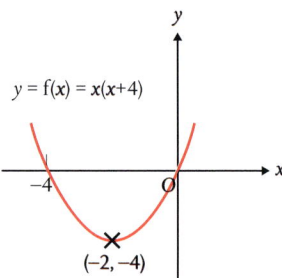

b

Reflect the graph in the y-axis.

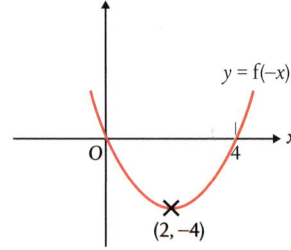

c

Reflect the graph in the x-axis.

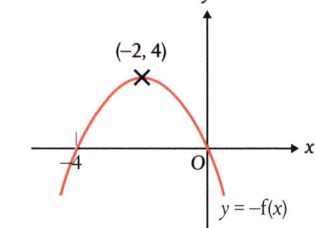

Chapter 4: More graphs

Practice

1 $f(x) = x^3$

Here is a sketch graph of $y = f(x)$.

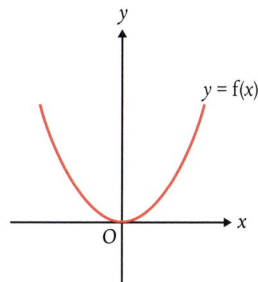

Copy the sketch graph.

On the same axes, sketch and label the graphs of

a $y = -f(x)$ **b** $y = f(x) + 2$ **c** $y = f(x - 3)$

2 $f(x) = x^2 + 3x$

Here is a sketch graph of $y = f(x)$.

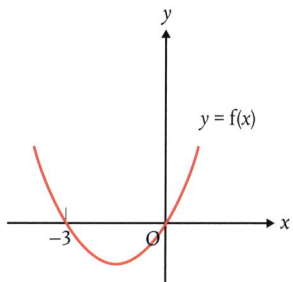

Copy the sketch graph.

On the same axes, sketch and label the graphs of

a $y = f(-x)$ **b** $y = f(x) - 1$ **c** $y = f(x + 2)$

3 $f(x) = x^3 + 4$

Here is a sketch graph of $y = f(x)$.

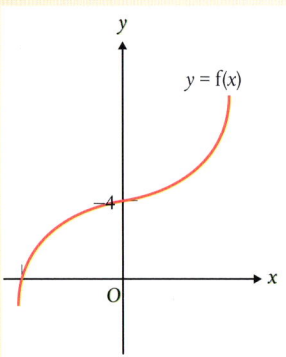

Copy the sketch graph.

On the same axes, sketch and label the graphs of

a $y = f(-x)$ **b** $y = f(x) - 6$ **c** $y = f(x + 3)$

4 $f(x) = \cos x$

Here is a sketch graph of $y = f(x)$ for $-360° \leq x \leq 360°$.

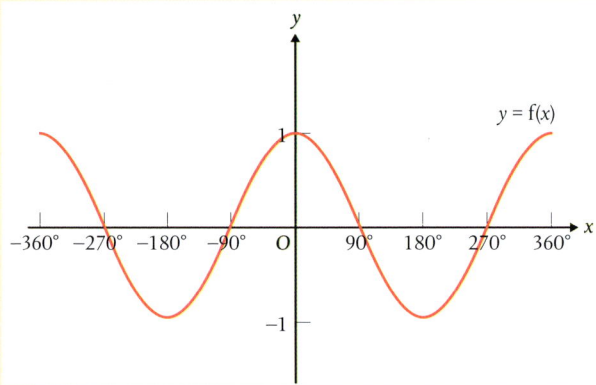

Copy the sketch graph.

On the same axes, sketch and label the graphs of

a $y = -f(x)$ **b** $y = f(x) + 1$ **c** $y = f(x - 90°)$

5 $f(x) = (x + 1)(x - 3)$

On the same axes, sketch and label the graphs of

a $y = f(x)$ **b** $y = f(x) - 2$ **c** $y = f(x + 3)$

6 $f(x) = x(x - 2)^2$

On the same axes, sketch and label the graphs of

a $y = f(x)$ **b** $y = -f(x)$ **c** $y = f(x - 2)$

7 f(x) = tan x

On the same axes, sketch and label each graph for $-360° \leq x \leq 360°$.

  a $y = f(x)$ **b** $y = f(-x)$ **c** $y = f(x) + 2$

> **Exam-style question**
>
> **8** $y = (x - 3)(x + 2)(x - 1)$
>
>   **a** Find the value of y when $x = 4$.   **(1 mark)**
>
>   **b** Write $y = (x - 3)(x + 2)(x - 1)$ in the form $y = ax^3 + bx^2 + cx + d$.   **(3 marks)**
>
>   **c** Sketch the graph of $y = (x - 3)(x + 2)(x - 1)$.
>
>   Show clearly the coordinates of any points of intersection of the graph with the axes.   **(3 marks)**
>
>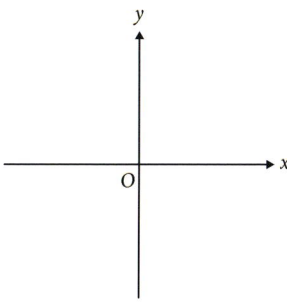
>
>   **d** Given that f(x) = $(x - 3)(x + 2)(x - 1)$, sketch the graph of $y = f(-x)$.
>
>   Show clearly the coordinates of any points of intersection of the graph with the axes.   **(2 marks)**

9 f(x) = x^2

On the same axes, sketch and label the graphs of

  a $y = f(x)$ **b** $y = -f(x) + 1$

Hint for Q9b

$y = -f(x) + 1$ is a reflection of the graph of $y = f(x)$ in the x-axis followed by a translation by the vector $\begin{pmatrix} 0 \\ 1 \end{pmatrix}$.

10 f(x) = $(x - 2)(x - 1)$

On the same axes, sketch and label the graphs of

  a $y = f(x)$ **b** $y = f(-x) - 2$ **c** $y = -f(x + 3)$

11 f(x) = sin x

On the same axes, sketch and label each graph for $-360° \leq x \leq 360°$.

  a $y = f(x)$ **b** $y = f(-x) + 1$ **c** $y = -f(x - 180°)$

Hint for Q10c

$y = -f(x + 3)$ is a translation of the graph of $y = f(x)$ by the vector $\begin{pmatrix} -3 \\ 0 \end{pmatrix}$ followed by a reflection in the x-axis.

 Talking point

What do you notice about the answer to **Q11c**? Explain why this happens.

12 $f(x) = x^3$

    a On the same axes, sketch and label the graphs of $y = f(x)$ and $y = f(x + 2)$.

    b The coordinates of the turning point of the graph of $y = f(x)$ are $(0, 0)$.

       Write the coordinates of the turning point of the graph of $y = f(x + 2)$.

13 $f(x) = x(x - 3)$

Here is the graph of $y = f(x)$.

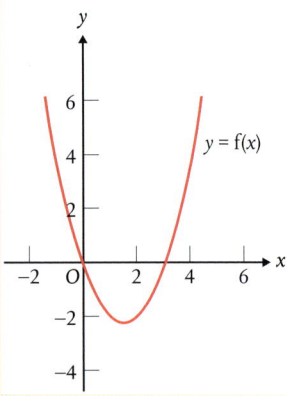

The roots of $f(x)$ are 0 and 3.

Find the roots of

  a $y = -f(x)$ **b** $y = f(-x)$ **c** $y = f(x + 1)$

14 $f(x) = x(x + 2)^2$

Here is the graph of $y = f(x)$.

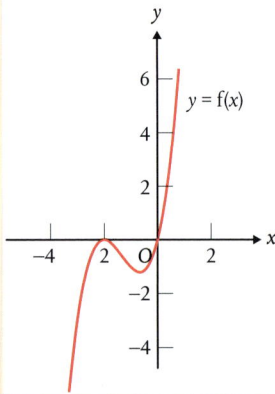

The roots of $f(x)$ are -2 and 0.

Find the roots of

  a $y = f(-x)$ **b** $y = f(x - 3)$ **c** $y = -f(x + 1)$

15 Problem-solving Here is the graph of $y = f(x)$.

P, Q and R are points on the graph of $y = f(x)$.

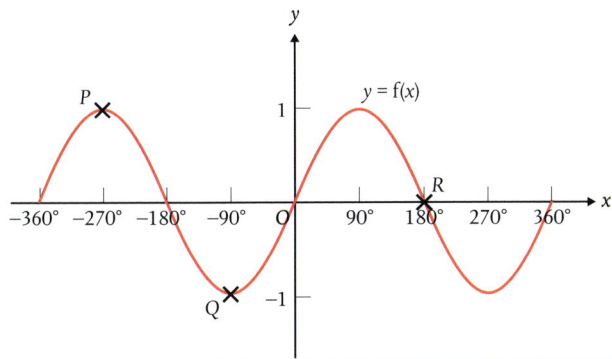

Find the coordinates of the points P, Q and R following each transformation.

a $f(x) - 3$
b $-f(x) + 1$
c $f(x - 90°) + 2$

16 Problem-solving Here is the graph of $y = f(x)$.

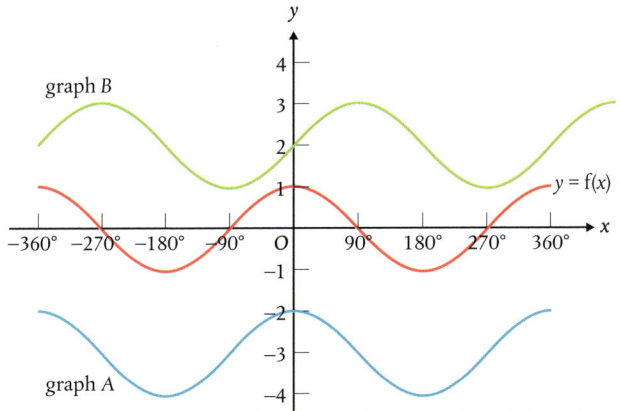

Graphs A and B are transformations of the graph of $y = f(x)$.

Describe each graph in terms of $f(x)$.

Chapter 4: More graphs

4.2 Stretching graphs

The graph of $y = af(x)$ is a stretch of the graph $y = f(x)$ by a scale factor of a in the vertical direction.

The graph of $y = f(ax)$ is a stretch of the graph $y = f(x)$ by a scale factor of $\frac{1}{a}$ in the horizontal direction.

Example 1

$f(x) = (x + 1)(x - 1)$

This is the graph of $y = f(x)$.

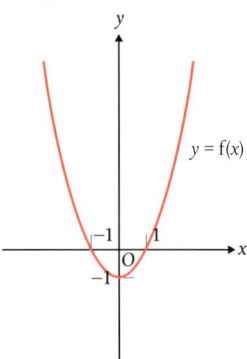

Sketch the graphs of

a $y = 3f(x)$ **b** $y = f(2x)$

a

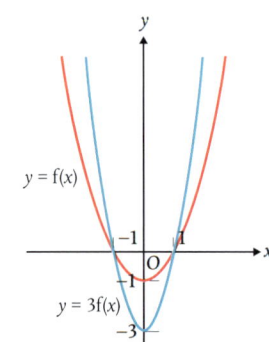

Stretch the graph by a scale factor of 3 in the vertical direction.

b

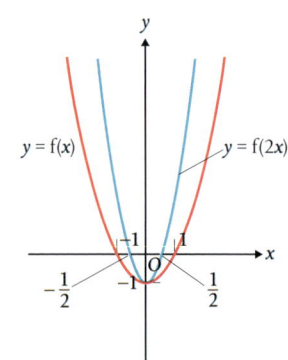

Stretch the graph by a scale factor of $\frac{1}{2}$ in the horizontal direction.

💬 Talking point

What happens to the coordinates for stretches in each direction?

Chapter 4: More graphs

Example 2

$f(x) = \sin x$

Sketch and label the graphs of $y = f(x)$ and $y = 2f(x - 90) + 3$.

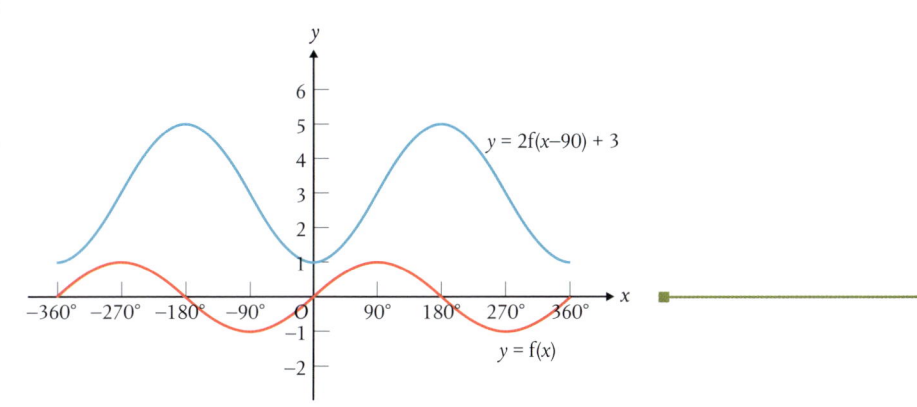

Translate the graph of $y = \sin x$ by $\begin{pmatrix} 90 \\ 0 \end{pmatrix}$, then stretch by a scale factor of 2 in the vertical direction and then translate by $\begin{pmatrix} 0 \\ 3 \end{pmatrix}$.

Practice

1 $f(x) = x(x - 1)$

Here is a sketch graph of $y = f(x)$.

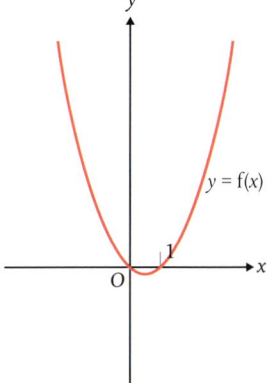

Copy the sketch graph.

On the same axes, sketch and label the graphs of

a $y = 2f(x)$ b $y = f(3x)$

Chapter 4: More graphs

2 f(x) = x(x + 2)²

Here is a sketch graph of y = f(x).

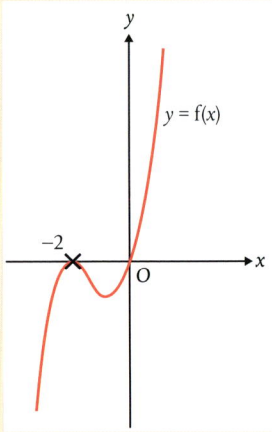

Copy the sketch graph.

On the same axes, sketch and label the graphs of

a y = 3f(x) **b** y = f($\frac{1}{2}$x)

Hint for Q2b

The reciprocal of $\frac{1}{2}$ is 2. Therefore, the graph of y = f($\frac{1}{2}$x) is a stretch of the graph y = f(x) by a scale factor of 2 in the horizontal direction.

3 f(x) = cos x

Here is a sketch graph of y = f(x).

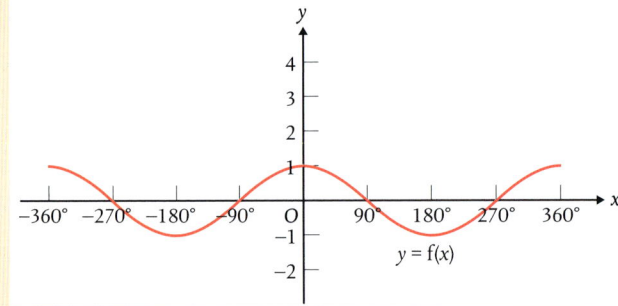

Copy the sketch graph.

On the same axes, sketch and label the graphs of

a y = 2f(x)

b y = f(3x)

c y = f($\frac{1}{2}$x)

Chapter 4: More graphs

4 f(x) = sin x

Here is a sketch graph of y = f(x).

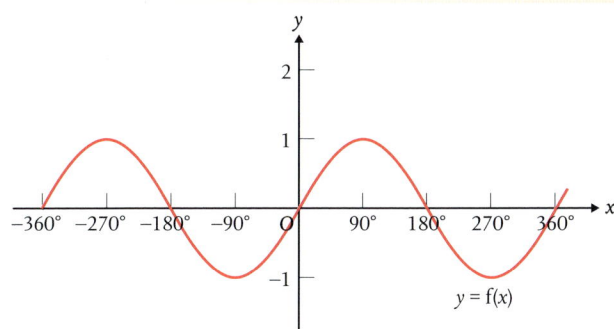

Copy the sketch graph.

On the same axes, sketch and label the graphs of

a $y = \frac{1}{2}f(x)$
b $y = f(2x)$

5 f(x) = (x + 2)(x − 1)

On the same axes, sketch and label the graphs of

a y = f(x)
b y = 2f(x)
c y = f(3x)

6 $f(x) = x(x − 1)^2$

On the same axes, sketch and label the graphs of

a y = f(x)
b $y = f(\frac{1}{2}x)$
c y = 3f(x)

7 f(x) = x(x + 4)

On the same axes, sketch and label the graphs of

a y = f(x)
b y = 2f(x) + 3

Hint for Q7b
y = 2f(x) + 3 is a stretch of the graph of y = f(x) by a scale factor of 2 in the vertical direction followed by a translation by the vector $\begin{pmatrix} 0 \\ 3 \end{pmatrix}$.

8 f(x) = x(x − 1)(x − 2)

On the same axes, sketch and label the graphs of

a y = f(x)
b y = f(3x) − 2
c y = −2f(x)

Hint for Q8c
y = −2f(x) is a stretch of the graph of y = f(x) by a scale factor of 2 in the vertical direction followed by a reflection in the x-axis.

9 f(x) = tan x

On the same axes, sketch and label the graphs of

a y = f(x)
b y = 2f(−x) + 1
c y = −3f(x + 90°)

10 $f(x) = x^2 − 4$

a On the same axes, sketch and label the graphs of y = f(x) and y = 2f(x).

b Work out the coordinates of the turning point of each graph.

Chapter 4: More graphs

11 $f(x) = x(x-3)(x+2)$

Here is the graph of $y = f(x)$.

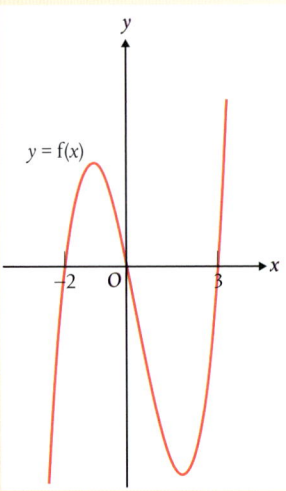

Find the roots of

a $y = 2f(x)$ **b** $y = f(2x)$

12 $f(x) = x(x-3)^2$

Work out the roots of

a $y = f\left(\frac{1}{2}x\right)$ **b** $y = \frac{1}{2}f(x)$

13 $f(x) = x(x-1)$

Work out the roots of

a $y = -2f(x)$ **b** $y = f(3x+1)$

14 Here is the graph of $y = f(x)$.

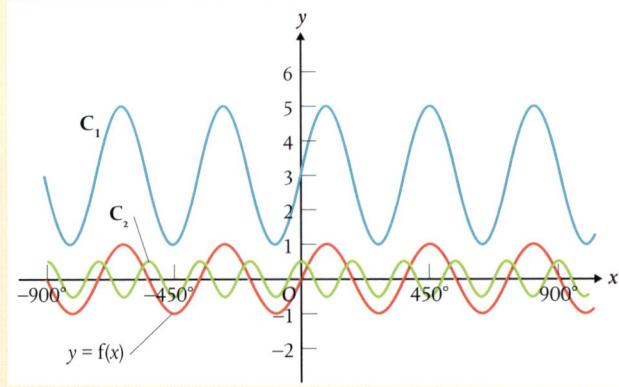

The graphs of C_1 and C_2 are transformations of the graph of $y = f(x)$.

Describe each graph in terms of $f(x)$.

Chapter 4: More graphs

Exam-style question

15 Here is the graph of $y = f(x)$.

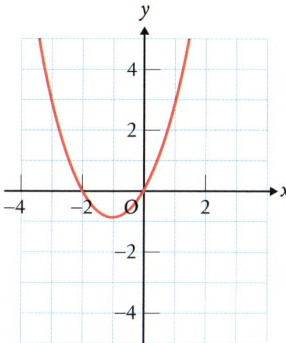

a Copy the grid below and draw the graph of $y = 2f(x)$.

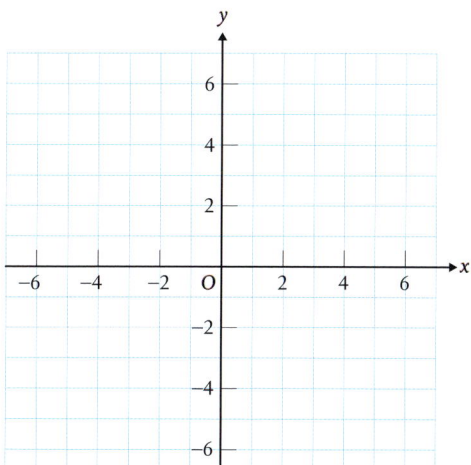

(2 marks)

b State the y-coordinates of the following points on the graph of $y = f(2x)$.

$(-3, \underline{}), (-2, \underline{}), (-1, \underline{}), (0, \underline{}), (1, \underline{})$ **(2 marks)**

16 **Problem-solving** The transformation $af(bx) - 1$ of the curve $f(x)$ moves point P from $(-2, 5)$ to $(1, 14)$.

Work out the values of a and b.

Chapter 4: More graphs

4.3 Circles

$x^2 + y^2 = r^2$ is the equation of a circle with centre (0, 0) and radius r.

$f(x - a)$ is a translation of the graph of $f(x)$ by $\begin{pmatrix} a \\ 0 \end{pmatrix}$.

$f(y - b)$ is a translation of the graph of $f(y)$ by $\begin{pmatrix} 0 \\ b \end{pmatrix}$.

Example 1

Sketch a graph of the circle $(x - 2)^2 + y^2 = 16$.

$x^2 + y^2 = 16$ is a circle centre (0, 0), radius 4.

$(x - 2)^2 + y^2 = 16$ is a translation of the circle centre (0, 0), radius 4 by $\begin{pmatrix} 2 \\ 0 \end{pmatrix}$.

$(x - 2)^2 + y^2 = 16$ is a circle centre (2, 0), radius 4.

$(x - 2)^2 + y^2 = 16 = f(x - 2)$

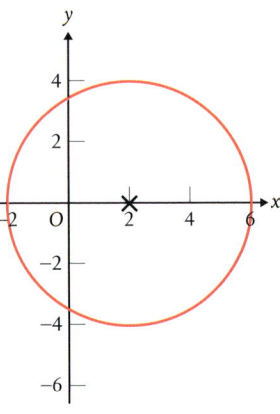

Example 2

Sketch a graph of the circle $x^2 + (y - 3)^2 = 16$.

$x^2 + y^2 = 16$ is a circle centre (0, 0), radius 4.

$x^2 + (y - 3)^2 = 16$ is a translation of the circle centre (0, 0), radius 2 by $\begin{pmatrix} 0 \\ 3 \end{pmatrix}$.

$x^2 + (y - 3)^2 = 4$ is a circle centre (0, 3), radius 4.

$x^2 + (y - 3)^2 = 4 = f(y - 3)$

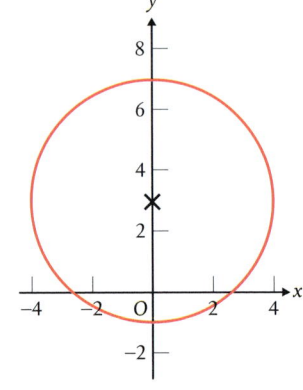

Chapter 4: More graphs

$(x-a)^2 + (y-b)^2 = r^2$ is the equation of a circle with centre (a, b) and radius r.

Example 3

Circle C has equation $x^2 - 4x + y^2 = 8$.

Work out the radius and the coordinates of the centre of circle C.

$x^2 - 4x + y^2 = 8$

$(x - 2^2) - 4 + y^2 = 8$ ← Complete the square for the x terms.

$(x - 2^2) + y^2 = 12$ ← Rearrange into the form $(x-a)^2 + y^2 = r^2$

Circle C has centre $(2, 0)$ and radius $\sqrt{12} = 2\sqrt{3}$

Practice

1. Sketch a graph of the circle $(x-1)^2 + y^2 = 9$.

2. Sketch a graph of the circle $x^2 + (y+2)^2 = 25$.

3. Find the equation of this circle.

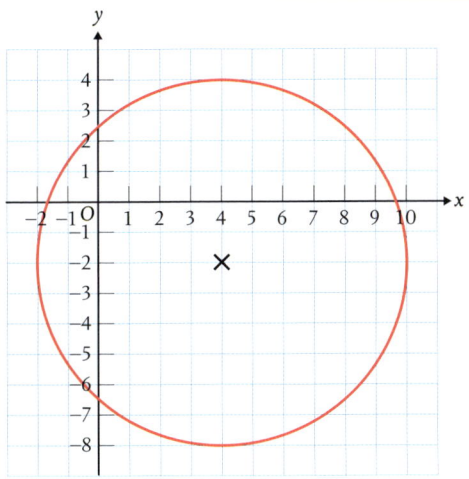

4. Sketch a graph of the circle $(x+2)^2 + (y-1)^2 = 4$.

Hint for Q1

Use compasses to draw the circle. Mark the centre with a cross.

Hint for Q2

This graph is a translation of the circle centre $(0, 0)$, radius $\sqrt{25}$ by $\begin{pmatrix} 0 \\ -2 \end{pmatrix}$.

5   a   Write down the coordinates of the centre of this circle.

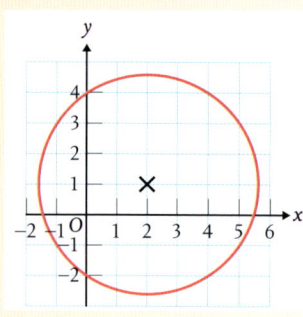

b   Use Pythagoras' theorem to find the distance from the centre to one of the y-intercepts. Leave the answer in surd form.

c   Hence, write the equation of the circle.

6   **Problem-solving** Write the equation of the locus of points 2.5 from $(-1, 7)$.

7   Here are some equations of circles. Work out the radius and the coordinates of the centre of each circle.

Hint for Q7c

Divide by a common factor first.

a $x^2 - 10x + y^2 = 5$ b $x^2 + 6x + y^2 = -2$ c $2x^2 - 8x + 2y^2 = 0$

8   Here are some equations of circles. Work out the radius and the coordinates of the centre of each circle.

a $x^2 + y^2 - 8y = 0$ b $x^2 + y^2 + 2y = 5$ c $3x^2 + 3y^2 + 12y = 39$

Hint for Q9

Complete the square for x and complete the square for y.

9   Here are some equations of circles. Work out the radius and the coordinates of the centre of each circle.

a $x^2 - 2x + y^2 + 6y = 10$ b $x^2 + 6x + y^2 - 4y = 0$ c $2x^2 + 4x + 2y^2 - 4y = 4$

Exam-style question

10 Circle **C** has equation $2x^2 + 2y^2 - 12x + 28y + 94 = 0$.

Work out the radius and the coordinates of the centre of circle **C**.   **(4 marks)**

Chapter 4: More graphs

11 a **Reasoning** Show that the circle with equation $x^2 + (y - 5)^2 = 10$ passes through the point (3, 6).

  b **Reasoning** Find the equation of the radius of the circle from part **a** that passes through the point (3, 6).

 Talking point
Explain how to answer a 'show that' question.

12 **Problem-solving** Find the equation of the radius of the circle $x^2 - 6x + y^2 - 8y = -20$ that passes through the point (5, 5).

Give the answer in the form $ax + by + c = 0$.

13 Find the equation of the tangent to the circle $(x - 6)^2 + (y - 1)^2 = 25$ at the point (10, 4).

Give the answer in the form $ax + by + c = 0$.

Hint for Q13
First find the gradient of the radius at the point (10, 4).

14 **Problem-solving** Find the equation of the tangent to the circle $x^2 + 8x + y^2 = 9$ at the point (−7, −4).

Give the answer in the form $ax + by + c = 0$.

15 **Reasoning** Show that the line $3y + 4x = 35$ is a tangent to the circle

$(x - 1)^2 + (y - 2)^2 = 25$ at the point (5, 5).

Chapter 4: More graphs

4.4 Exponential and reciprocal graphs

Exponential functions are functions of the form $f(x) = a^x$, where $a > 0$.

The graph of an exponential function has one of these shapes.

Exponential growth **Exponential decay**

$y = a^x$, where $a > 1$ or $y = b^{-x}$, where $0 < b < 1$ $y = a^{-x}$, where $a > 1$ or $y = b^x$, where $0 < b < 1$

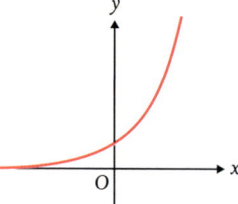 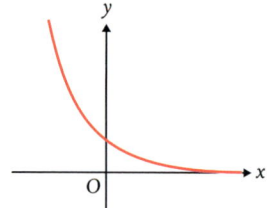

The graphs of reciprocal functions of the form $y = \dfrac{a}{x}$ and $y = \dfrac{a}{x^2}$ have asymptotes at $x = 0$ and $y = 0$.

An asymptote is a line that the graph approaches but never reaches.

$y = \dfrac{a}{x}$, where $a > 0$ $y = \dfrac{a}{x^2}$, where $a > 0$

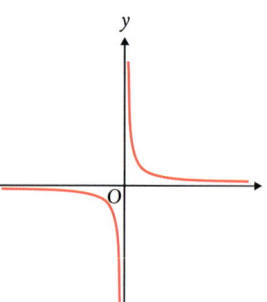 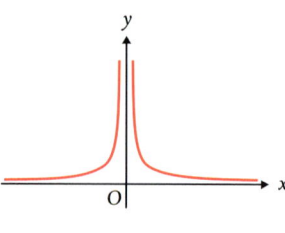

> **Talking point**
> Describe the similarities and differences between the graphs of the reciprocal functions.

$y = \dfrac{a}{x}$, where $a < 0$ $y = \dfrac{a}{x^2}$, where $a < 0$

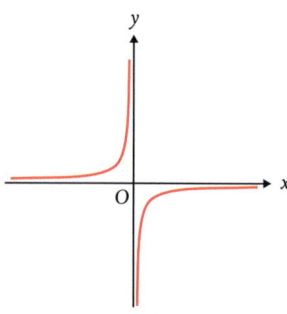 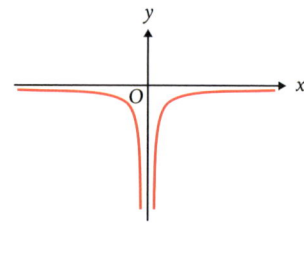

Chapter 4: More graphs

Example 1

$f(x) = a^x$

This is the graph of $y = f(x)$.

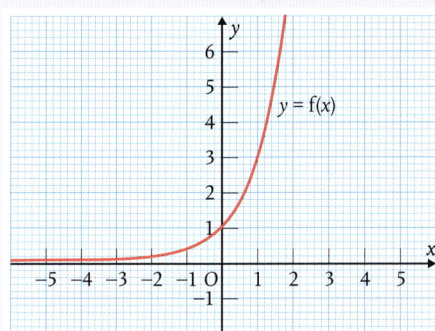

Work out the value of a.

Using the graph,

$x = 1$ when $y = 3$ ◀ — Use the coordinates of a point on the graph.

Substituting these values into $y = a^x$ gives:

$3 = a^1$.

Therefore, $a = 3$.

Example 2

Sketch and label the graph

a $y = \dfrac{2}{x}$ b $y = -\dfrac{4}{x^2}$

a

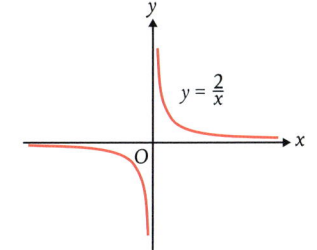

This is a $y = \dfrac{a}{x}$ graph with $a > 0$.

b

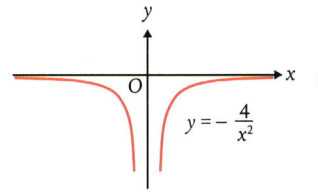

This is a $y = \dfrac{a}{x^2}$ graph with $a < 0$.

4.4 Exponential and reciprocal graphs

73

Practice

1 a On the same axes, sketch and label the graphs $y = 1.5^x, y = 2^x, y = 3^x$ and $y = 0.5^x$.

b Use the sketches to describe the transformation of $y = 2^x$ to $y = 0.5^x$.

2 a On graph paper, draw the graph of $y = 0.8^x$, for $-4 \leq x \leq 4$.

b Use the graph to solve the equation $0.8^x = 2$.

3 $f(x) = a^x$

This is the graph of $y = f(x)$.

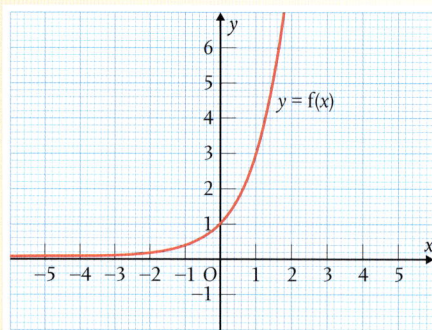

Work out the value of a.

4 $f(x) = a^{-x}$

This is the graph of $y = f(x)$.

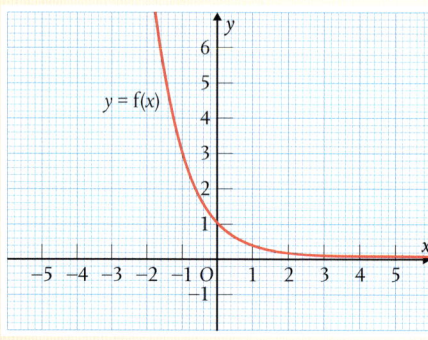

Work out the value of a.

Chapter 4: More graphs

Exam-style question

5 Copy the axes below and sketch the graph of $y = 2^x$.

State clearly any asymptote and the coordinates of any points of intersection of the graph with the axes.

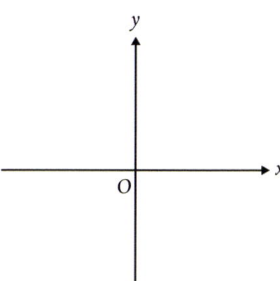

(3 marks)

6 $f(x) = 3^x$

One the same axes, sketch the graphs of

   a $y = f(x)$ **b** $y = f(x) - 3$ **c** $y = 3f(x)$

7 **Problem-solving** The graph of $y = ab^x$ passes through the points $(1, 0.8)$ and $(-2, 12.5)$.

By drawing a sketch or otherwise, explain why $0 < b < 1$.

8 On the same axes, sketch and label the graphs

   a $y = \frac{3}{x}$ **b** $y = -\frac{3}{x}$ **c** $y = \frac{7}{x}$

> **Talking point**
> Describe the similarities and differences between the graphs of the reciprocal functions in **Q8**.

9 On the same axes, sketch and label the graphs.

   a $y = \frac{2}{x^2}$ **b** $y = -\frac{2}{x^2}$ **c** $y = -\frac{7}{x^2}$

10 $f(x) = \frac{2}{x}$

On the same axes, sketch each graph and write the equations of the asymptotes.

   a $y = f(x)$ **b** $y = f(x) - 1$ **c** $y = 3f(x)$

11 $f(x) = \frac{3}{x^2}$

On the same axes, sketch each graph and write the equations of the asymptotes.

   a $y = f(x)$ **b** $y = f(x) + 2$ **c** $y = f(x - 1)$

12 $f(x) = -\frac{2}{x}$

The graph of $y = f(x)$ is transformed. Write the equations of the asymptotes for each transformation.

   a $y = -f(x)$ **b** $y = f(x) + 1$ **c** $y = f(x + 2)$

Chapter 4: More graphs

13 $f(x) = \frac{1}{x^2}$

The graph of $y = f(x)$ is transformed. Write the equations of the asymptotes for each transformation.

    a $y = f(x) - 5$ b $y = f(3x)$ c $y = f(x + 4)$

14 **Reasoning** $f(x) = \frac{1}{x}$

Hint for Q14

Apply the transformations to the graph of $y = \frac{1}{x}$.

Here are some transformations of the graph of $y = f(x)$.

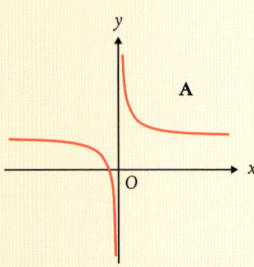
A

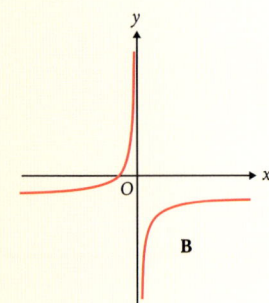
B

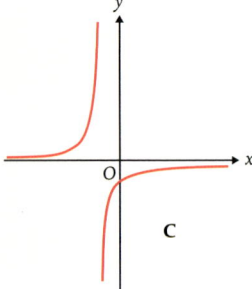
C

Match each graph to its transformation.

    i $y = f(x) + 2$ ii $y = -f(x + 1)$ iii $y = -f(x) - 2$

Exam-style question

15 Copy the axes below, and sketch the graph of $y = \frac{1}{x} - 2$.

Show clearly any asymptote and the coordinates of any points of intersection of the graph with the axes.

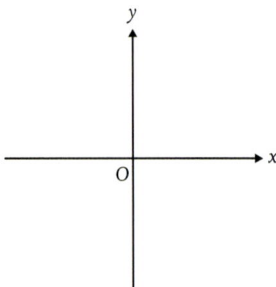

(4 marks)

16 **Reasoning** $f(x) = ab^x$

This is the graph of $y = f(x)$.

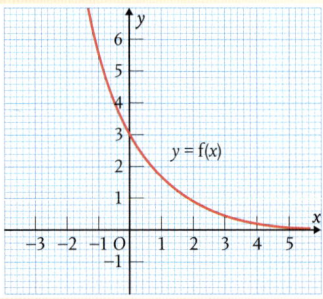

Work out the values of a and b.

Chapter 4: More graphs

4.5 Non-linear graphs

The gradient of a curve is constantly changing. Use the tangent at a point on the curve to find the gradient of the curve at that point. The tangent at a point to a curve is the straight line that just touches, but does not cross, the curve at that point.

Example 1

This is the graph of $y = -\frac{x}{2}(x - 10)$.

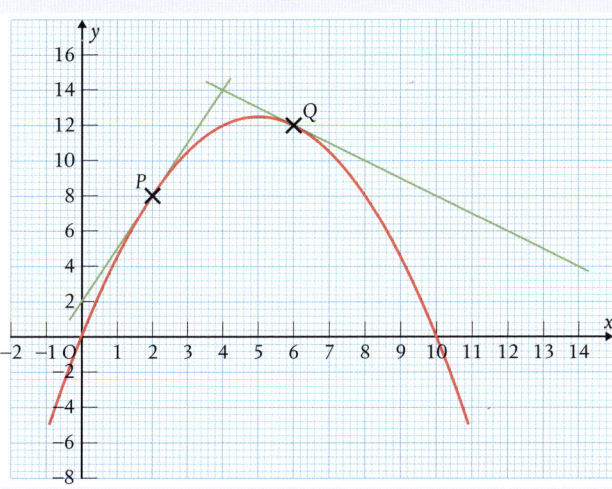

Points P and Q are on the curve.

Estimate the gradient at each point.

Gradient at point P:

$$m = \frac{14 - 2}{4 - 0} = \frac{12}{4} = 3$$

gradient $= \frac{y_2 - y_1}{x_2 - x_1}$

Gradient at point Q:

$$m = \frac{14 - 4}{4 - 14} = \frac{10}{-10} = -1$$

To estimate the area under a part of a curved graph, split the area under the curve into a number of trapeziums of equal width. This method is called the trapezium rule.

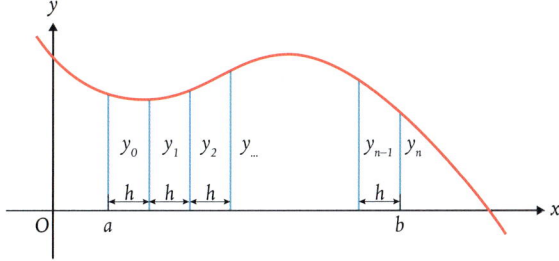

The trapezium rule is:

$$\text{area} = \tfrac{1}{2}h[y_0 + 2(y_1 + y_2 + \ldots + y_{n-1}) + y_n]$$

where h is the width of each strip, $y_0, y_1, y_2, \ldots, y_{n-1}, y_n$ are the values of y for each value of x used, and n is the number of equal strips the area has been divided up into; $x = a$ and $x = b$ define the vertical boundaries of the area.

Chapter 4: More graphs

The number of x-values and the number of y-values is one more than the number of strips n.

Use $h = \dfrac{b-a}{n}$ to calculate the width of each strip, h.

Example 2

This is the graph of $y = -\dfrac{x}{5}(x-10)$.

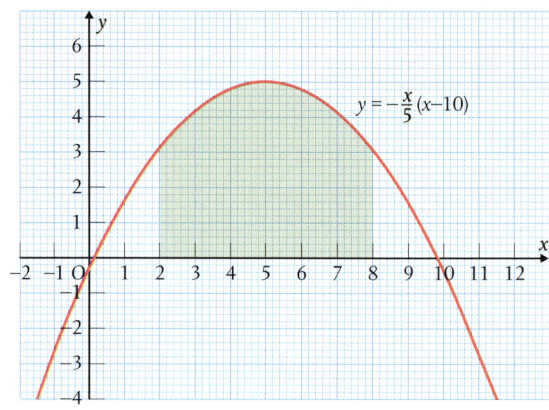

Use the trapezium rule to find an estimate of the area of the region under the curve with equation $y = -\dfrac{x}{5}(x-10)$, between $x = 2$, $x = 8$ and the x-axis. Use three strips of equal width.

$h = \dfrac{b-a}{n} = \dfrac{8-2}{3} = 2$ ── Each strip will have a width of 2.

Use a table to record y for each value of x.

x	2	4	6	8
$y = -\dfrac{x}{5}(x-10)$	3.2	4.8	4.8	3.2

From the table,
$y_0 = 3.2, y_1 = 4.8, y_2 = 4.8, y_3 = 3.2$

$\text{area} = \dfrac{1}{2}h[y_0 + 2(y_1 + y_2) + y_3]$

$= \dfrac{1}{2} \times 2[3.2 + 2(4.8 + 4.8) + 3.2]$

$= 3.2 + 2(9.6) + 3.2$

$= 25.6$ square units

Practice

1. This is the graph of $y = 0.5x^2 + x - 1$.

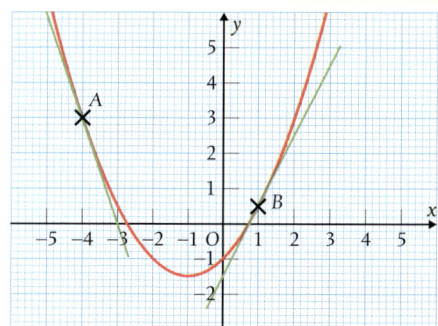

   Points A and B are on the curve.

   Estimate the gradient at each point.

2. This is the graph of $y = x^3 + 4x^2 + x - 1$.

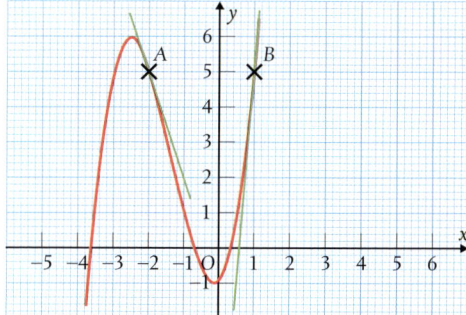

   Points A and B are on the curve.

   Estimate the gradient at each point.

3. a On graph paper, draw the graph of $y = -x^2 + x + 4$, for $-2 \leq x \leq 3$.

   b Use the graph to estimate the gradient of the curve at $x = 1$.

   c Use the graph to estimate the gradient of the curve at $x = -1$.

4 **Reasoning** The distance–time graph shows information about the start of a car journey.

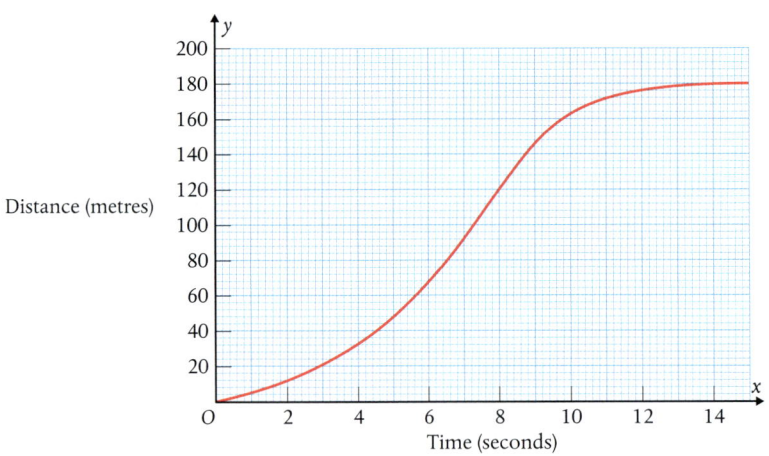

a Use the graph to estimate the speed at 8 seconds into the journey.

b Describe the change in speed over the 15 seconds.

5 **Reasoning** The velocity–time graph describes the motion of a car.

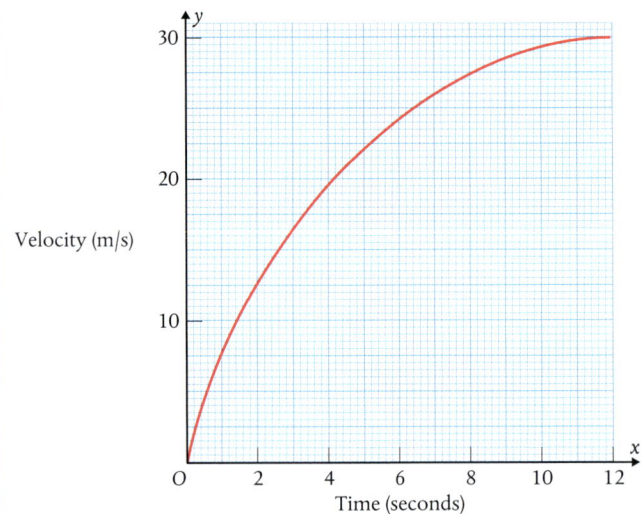

a Estimate the gradient at time $t = 6$ seconds.

b What does the gradient represent?

Chapter 4: More graphs

6 Use the trapezium rule to find an estimate of the area of the shaded region using five strips of equal width.

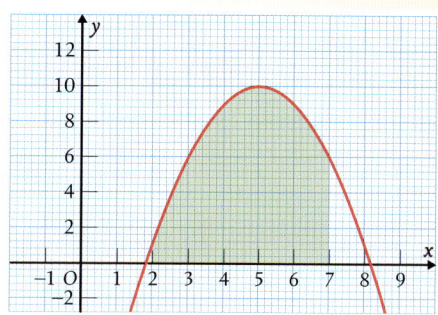

7 Use the trapezium rule to find an estimate of the area of the shaded region using four strips of equal width.

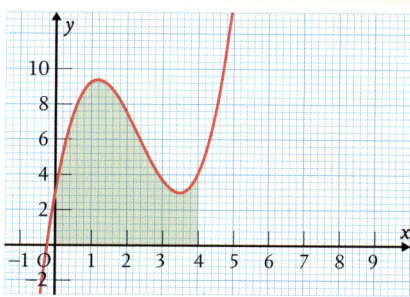

8 a Copy and complete the table of values for $y = -x^2 + 3x + 10$.

x	1	2	3	4
y				

   b Use the trapezium rule to find an estimate of the area of the region between the curve with equation $y = -x^2 + 3x + 10$ and the x-axis from $x = 1$ to $x = 4$.

   Use three strips of equal width.

Exam-style question

9 a Copy and complete the table of values for $y = 3^x$.

x	0	1	2	3	4
y					

   (2 marks)

   b Use the trapezium rule to find an estimate of the area of the region under the curve with equation $y = 3^x$, between $x = 0$, $x = 4$ and the x-axis.

   Use four strips of equal width. **(3 marks)**

Chapter 4: More graphs

Hint for Q10
Draw a table to record y for each value of x.

10 Use the trapezium rule to find an estimate of the area of the region between the curve with equation $y = -0.2x^2 + 0.6x + 8$ and the x-axis, from $x = 0$ to $x = 6$.

Use three strips of equal width.

11 Use the trapezium rule to find an estimate of the area of the region between the curve with equation $y = -x^3 + 11x^2 - 36x + 39$ and the x-axis, from $x = 1$ to $x = 6$.

Use five strips of equal width.

12 **Problem-solving** Use the trapezium rule to find an estimate of the area of the shaded region. Use four strips of equal width.

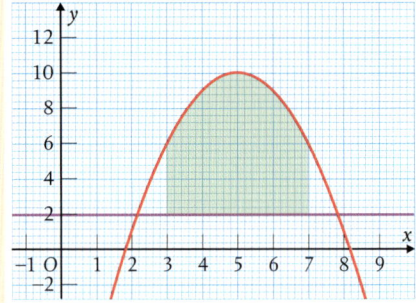

 Talking point
Explain how to work out the height of each trapezium when the area is not enclosed by the x-axis, as in Q12 and Q13.

13 **Problem-solving** Use the trapezium rule to find an estimate of the area of the shaded region. Use four strips of equal width.

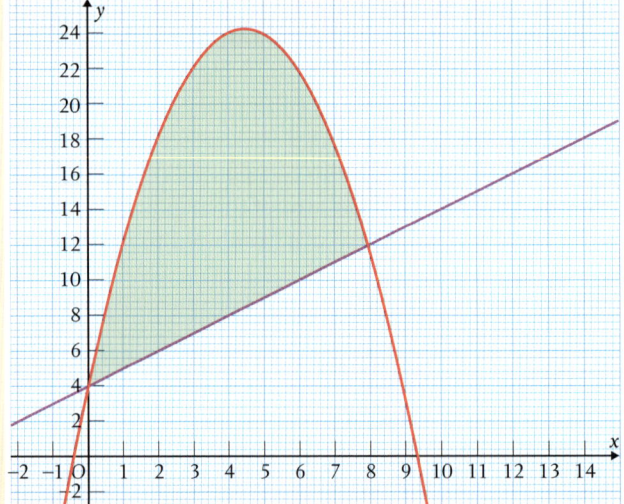

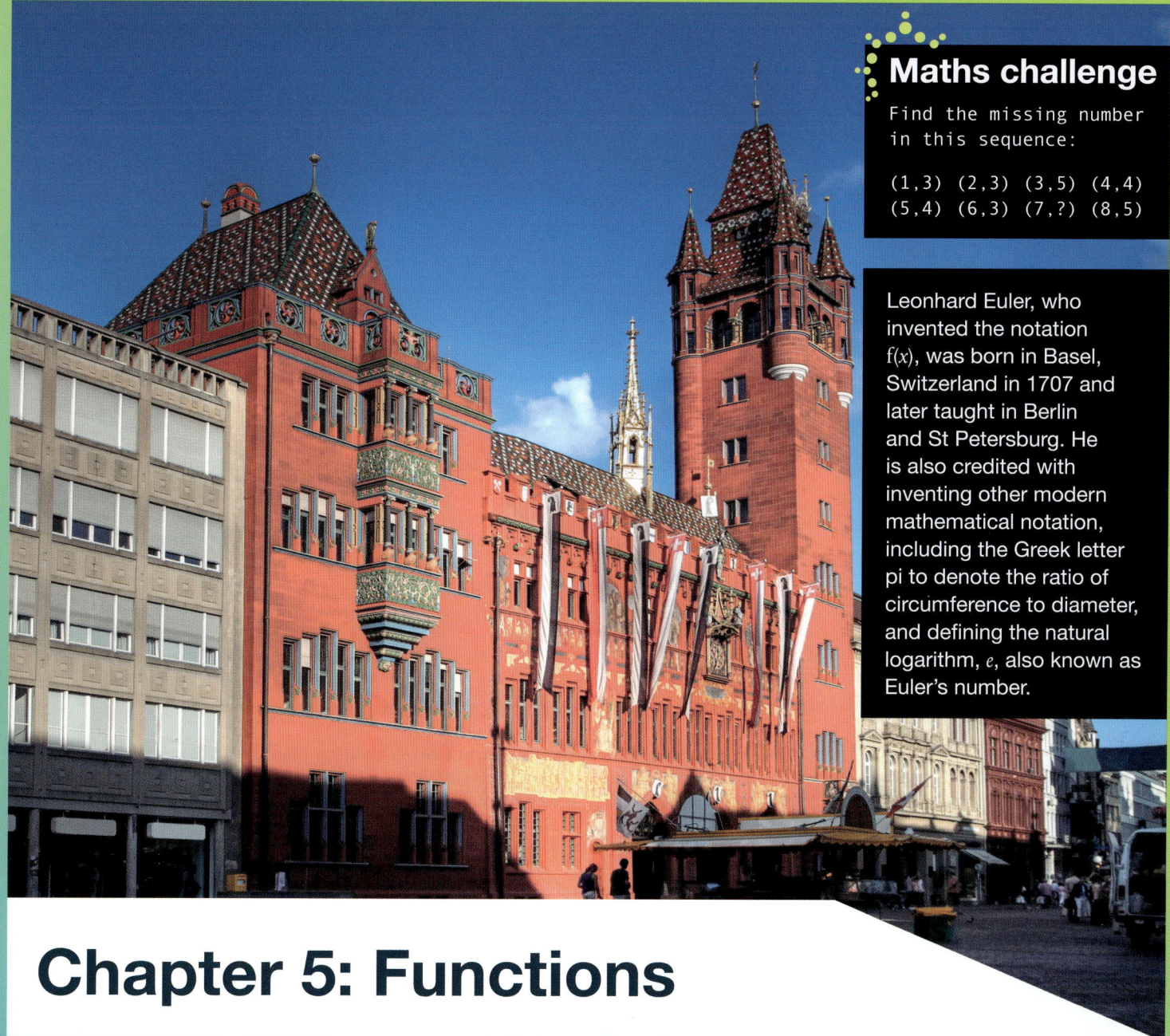

Maths challenge

Find the missing number in this sequence:

(1,3) (2,3) (3,5) (4,4)
(5,4) (6,3) (7,?) (8,5)

Leonhard Euler, who invented the notation f(x), was born in Basel, Switzerland in 1707 and later taught in Berlin and St Petersburg. He is also credited with inventing other modern mathematical notation, including the Greek letter pi to denote the ratio of circumference to diameter, and defining the natural logarithm, e, also known as Euler's number.

Chapter 5: Functions

In this chapter you will:

- identify the domain and range of a function from its graph and from its equation
- identify values that must be excluded from the domain of a function
- understand and use composite functions and sketch their graphs
- find the domain and range of a composite function
- find the inverse of a function
- transform functions and write the algebraic equation of transformed functions

Prior knowledge

- use function notation $f(x)$
- use set builder notation $\{x: \ 5 \leq x \leq 3\}$
- find the minimum point of a quadratic curve
- sketch linear, quadratic, cubic, trigonometric and exponential graphs
- sketch transformations of graphs
- apply transformations to graphs

Chapter 5: Functions

5.1 Functions

In mathematics, a **relation** is a link or connection between two sets.

These mapping diagrams shows relations.

'multiply by 4'

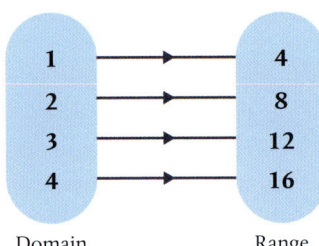

Domain → Range

'square'

Domain → Range

'is a factor of'

Domain → Range

Each element in the domain maps to exactly one element in the range.

Each input maps to only one output.

This is a **one-to-one** relation.

Each element in the domain maps to exactly one element in the range.

Each input maps to only one output.

Some outputs have more than one input.

This is a **many-to-one** relation.

Some elements in the domain map to more than one element in the range.

Each input **does not map** to only one output.

This is a **one-to-many** relation.

The **domain** is the set of input values that map to output values in the **range**.

Some relations are functions. For a **function**, each input maps to only one output.

One-to-one relations and many-to-one relations are functions.

One-to-many relations are not functions.

These mapping diagrams represent functions:

Read the notation f(x) as 'f of x'.

$f(x) = 4x$

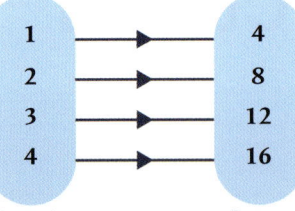

Domain → Range

$f(x) = x^2$

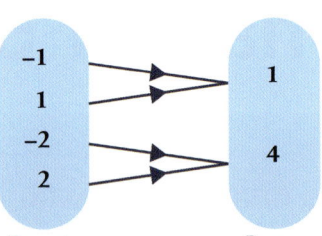

Domain → Range

$\mathbb{Z}$ is the set of all integers. {..., −3, −2, −1, 0, 1, 2, 3, 4, ...}

$\mathbb{R}$ is the set of real numbers. This includes

- integers
- rational numbers (numbers that can be written as fractions)
- irrational numbers (numbers that cannot be written as fractions, for example π).

Chapter 5: Functions

Example 1

a Sketch the graph of $y = 2x + 3$ for values of x between -5 and $+3$.

b Find the domain.

c Find the range.

a When $x = -5$, $y = 2x + 3 = 2 \times -5 + 3 = -7$ — Find the y-values for the minimum and maximum x-values.

When $x = 3$, $y = 2x + 3 = 2 \times 3 + 3 = 9$

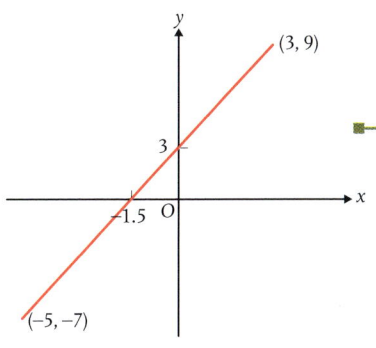

Label the x- and y-intercepts and the coordinates of the end points.

b Domain is $\{x \in \mathbb{R} : -5 \leq x \leq 3\}$ — The domain is the x-values plotted on the graph.

c Range is $\{f(x) \in \mathbb{R} : -7 \leq f(x) \leq 9\}$ — The range is the y-values plotted on the graph.

Example 2

Here is the graph of the function $f(x) = x^2$

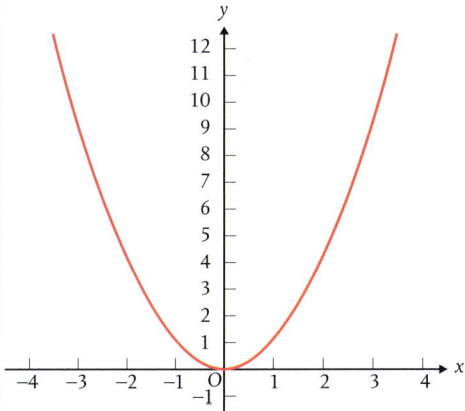

Find the domain and range of $f(x) = x^2$

Domain is $\{x \in \mathbb{R}\}$ — x can take any value, negative, positive, rational or irrational.

Range is $\{f(x) \in \mathbb{R} : f(x) \geq 0\}$ — $f(x)$ can take only positive real values (or zero).

Chapter 5: Functions

If a graph does not have a restricted domain or range, then the domain is $\{x \in \mathbb{R}\}$ and the range is $\{y \in \mathbb{R}\}$.

Any value of x that gives denominator 0 is excluded from the domain, because $\frac{1}{0}$ is undefined.

Here is the graph of $f(x) = \frac{1}{x}$.

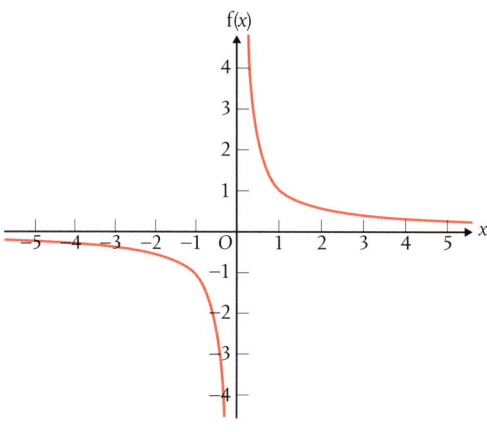

$x = 0$ is an asymptote. x cannot take the value 0 because $\frac{1}{0}$ is undefined.

Domain is $\{x \in \mathbb{R} : x \neq 0\}$

$f(x) = 0$ is an asymptote. $f(x)$ cannot take the value 0.

The range is $\{f(x) \in \mathbb{R} : f(x) \neq 0\}$

Example 3

Here is the graph of $f(x) = \dfrac{1}{x - 2}$

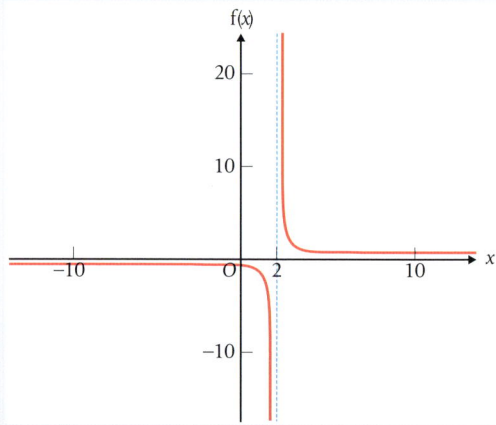

Find the domain and range of $f(x) = \dfrac{1}{x - 2}$.

$x = 2$ is an asymptote. x cannot take the value 2 because $\frac{1}{0}$ is undefined.

Domain is $\{x \in \mathbb{R} : x \neq 2\}$

$f(x) = 0$ is an asymptote. $f(x)$ cannot take the value 0.

Range is $\{f(x) \in \mathbb{R} : f(x) \neq 0\}$

Chapter 5: Functions

Practice

1 **Reasoning** For each mapping diagram, state whether the relation is a function.

For the functions, write the function f(x).

a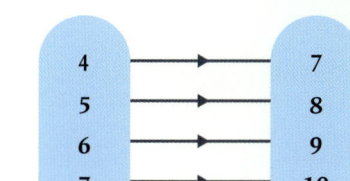

b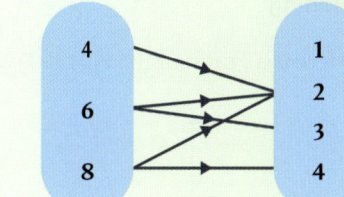

c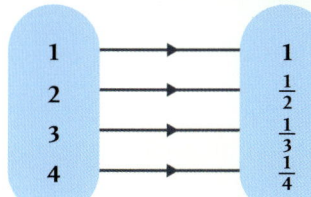

d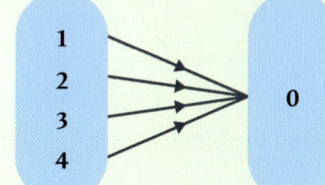

e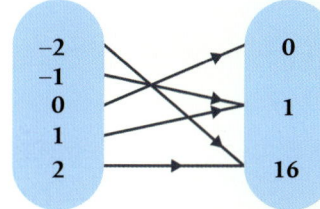

f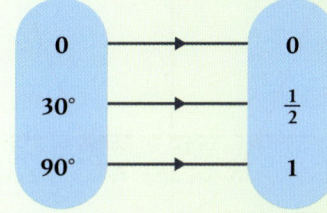

2 A graph passes through these points

(0, 0) (1, 1) (1, −1) (4, 2) (4, −2) (9, 3) (9, −3)

Hint for Q2
The domain is the x-values.

a Draw a mapping diagram to show the domain and range of this relation.

b **Reasoning** Is this relation a function? Explain.

3 $f(x) = x^2 - 1$ and $g(x) = 4x$
Work out

a f(3)
b g(−2)
c f(4) × g(1)
d $\dfrac{f(-5)}{g(2)}$
e f(7) − g(−1)
f 2f(6)

4 $f(x) = x^2 - 5x + 6$
Work out the values of a when f(a) = 0

5 $g(x) = x^2 + 6x - 3$
Work out the values of a when g(a) = −8

6 $h(x) = x^2 + 4x - 7$
Work out the values of a when h(a) = 11

7   a   Sketch the graph of $y = 3x - 2$ for values of x between -4 and $+4$.

    b   Find the domain.

    c   Find the range.

8   **Reasoning**

    a   Here is a graph of $f(x) = (x + 1)^2$ on the domain $\{x: -4 \leq x \leq 3\}$

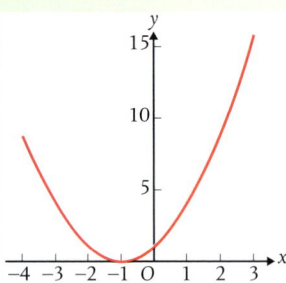

        Find the range.

    b   Here is a graph of $f(x) = x^2 - 6$ on the domain $\{x \in \mathbb{R}\}$

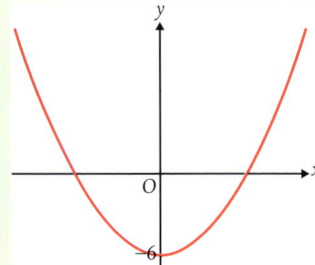

        Find the range.

9   **Reasoning** Here is a graph of $f(x) = -\frac{2}{x}$

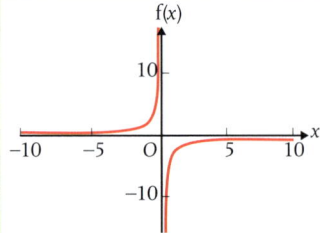

    State the domain and the range.

10 Here is a graph of $f(x) = \tan x$ for positive values of x less than $360°$.

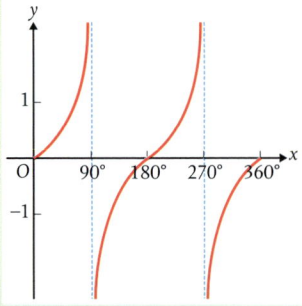

a State the domain.

Copy and complete

$\{x : 0 \leq x < \_\_\_ \cup 90° < x < \_\_\_ \cup 270° < x \leq \_\_\_\}$

b State the range.

11 Problem-solving Find the range of each function

a $y = 4x - 1$ on the domain $\{x: 0 \leq x \leq 6\}$

b $5x + 3y = 6$ on the domain $\{x: -3 \leq x \leq 3\}$

c $y - 3 = 2(x - 1)$ on the domain $x \in \mathbb{R}$.

12 Reasoning The domain of this function is $x \in \mathbb{Z}$.

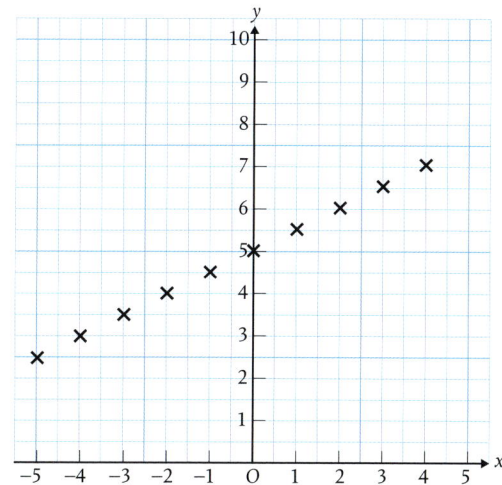

Hint for Q12

Write the coordinates of the points shown and decide on the possible values of y.

a Explain why there are no lines joining the points.

b Show that the range of this function is $f(x) \in \mathbb{R}$.

13 Find the range of each function

a $f(x) = x^2 + 4x - 3$

b $f(x) = 2x^2 + x + 5$

Hint for Q13

Find the minimum point of the curve: this is the minimum value of $f(x)$.

Exam-style question

14 Given that
$g(x) = 2x^2 + 4x - 5$
work out the range of $g(x)$. **(3 marks)**

15 Find the range of the function $f(x) = x^3 + 2x + 1$ on the domain $\{x: -3 \leq x \leq 3\}$

Hint for Q15

Visualise or sketch the graph.

16 Problem-solving Find the range of

a $f(x) = \cos x°$

b $f(x) = \sin(x + 30)°$

c $f(x) = 2\sin x°$

d $f(x) = \cos 2x°$

Talking point

How does visualising the graph help find the range?

17 This piecewise linear graph shows parking charges in a car park.

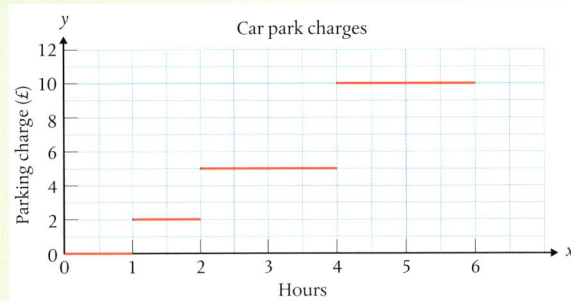

Copy and complete the domain for this function.

$y = 0$ $0 \leq x < \_\_\_\_$ $y = 2$ $1 \leq x < \_\_\_\_$

$y = 5$ $\_\_\_\_ \leq x < 4$ $y = 10$ $4 \leq x < \_\_\_\_$

18 This piecewise linear graph shows charges for plumbing repairs.

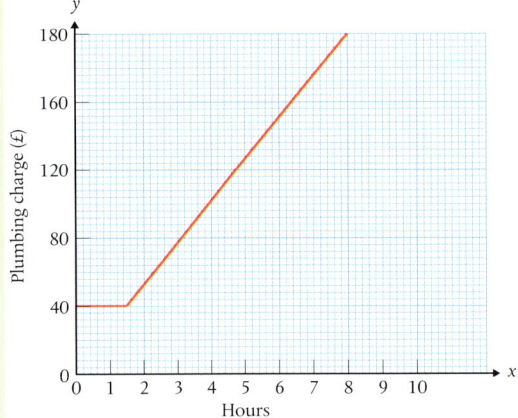

Copy and complete the domain for this function.

$y = 40$ $\_\_\_\_ < x \leq \_\_\_\_$ $y = 20x + 10$ $\_\_\_\_ < x \leq \_\_\_\_$

19 **Problem-solving** Find the domain of each function.

a $f(x) = \dfrac{3}{x}$

b $f(x) = \dfrac{3}{x - 4}$

c $f(x) = \dfrac{x}{(x - 3)(x + 2)}$

d $f(x) = \dfrac{1}{x^2 - 25}$

Exam-style question

20 Given that
$$f(x) = \dfrac{4}{x - 7}$$
write down the value of x that must be excluded from the domain of f(x). **(1 mark)**

Chapter 5: Functions

5.2 Composite functions

f(x) and g(x) are functions.

fg(x) is a **composite function**.

To work out fg(x), first work out g(x), and then substitute the answer into f(x).

Example 1

$f(x) = 3x^2$ and $g(x) = 2x + 1$

Work out

a fg(4)   **b** gf(−1)   **c** fg(0)

a $g(4) = 2 \times 4 + 1 = 9$ — First work out g(4).

$f(9) = 3 \times 9^2 = 243$ — Substitute the value of g(4).

b $f(-1) = 3 \times 1 = 3$ — First work out f(−1).

$g(3) = 2 \times 3 + 1 = 7$

c $g(0) = 2 \times 0 + 1 = 1$ — First work out g(0).

$f(1) = 3 \times 1^2 = 3$

Example 2

f and g are functions such that
$f(x) = 4x$
and
$g(x) = x^2 + 2$

a Write an expression for

    i fg(x)   **ii** gf(x)

b Sketch the graph of fg(x).

a i $fg(x) = f(x^2 + 2) = 4(x^2 + 2)$ — fg(x) = f(g(x))

  ii $gf(x) = g(4x) = ((4x)^2 + 2)$

              $= 16x^2 + 2$

              $= 2(8x^2 + 1)$

Chapter 5: Functions

b $fg(x) = 4(x^2 + 2)$

$= 4x^2 + 8$

Compare it to the graph of $y = x^2$ — This is a translation of the graph $y = x^2$ by $\begin{pmatrix} 0 \\ 8 \end{pmatrix}$ and a stretch scale factor 4 in the y-direction.

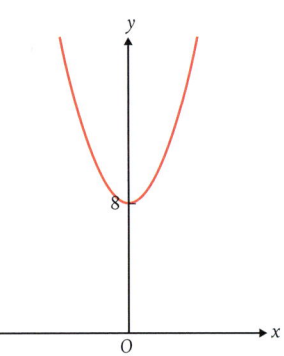

Practice

1 f and g are functions such that
$f(x) = x + 1$
and
$g(x) = x^2$

Find

  a $gf(2)$ **b** $fg(-4)$ **c** $gf(10)$

2 g and h are functions such that
$g(x) = \dfrac{3}{x - 2}$
and
$h(x) = x^2 + 4$

  a Find

    i $gh(0)$ **ii** $gh(5)$ **iii** $hg(1)$

  b **Reasoning** Show that $hg(2)$ is undefined.

3 f and g are functions such that
$f(x) = 2x - 1$
and
$g(x) = x + 4$

  a Write an expression for

    i $fg(x)$ **ii** $gf(x)$

  b Sketch the graph of $fg(x)$.

Chapter 5: Functions

4 g and h are functions such that
$g(x) = x + 3$
and
$h(x) = x^2 + 4x$

    a Write an expression for

        i $gh(x)$ **ii** $hg(x)$

    b Sketch the graph of $gh(x)$.

5 p and q are functions such that
$p(x) = x^3$
and
$q(x) = x - 1$

    a Write an expression for

        i $pq(x)$ **ii** $qp(x)$

    b Sketch the graph of

        i $pq(x)$ **ii** $qp(x)$

> **Hint for Q5b**
>
> Use transformations on the graph of $y = x^3$

Exam-style question

6 The functions $g(x)$ and $h(x)$ are defined as
$g(x) = 2x^2 - 4x - 5$
$h(x) = x - 2$

Given that
$j(x) = gh(x) + 7$
show that $j(x)$ can be written in the form $a(x + b)^2$, where a and b are integers. **(4 marks)**

7 **Problem-solving** f and g are functions such that
$g(x) = 3^x$
and
$h(x) = 2x + 1$

    a Write an expression for

        i $gh(x)$ **ii** $hg(x)$

    b Sketch the graph of

        i $gh(x)$ **ii** $hg(x)$

Chapter 5: Functions

8 **Problem-solving** f and g are functions such that
$f(x) = \cos x°$
and
$g(x) = x - 30°$

    a Write an expression for fg(x).

    b State the range of fg(x).

9 **Problem-solving** h and j are functions such that
$h(x) = 3x$
and
$j(x) = \sin x°$

    a Write an expression for jh(x) and state the range of the function.

    b Write an expression for hj(x) and state the range of the function.

10 **Reasoning** f and g are functions such that
$f(x) = \tan x°$
and
$g(x) = 2x$

    Show that fg(x) = is undefined for $x = 45°$

11 s and t are functions such that
$s(x) = \dfrac{x}{x^2 - 9}$
and
$t(x) = 2x + 1$

Find the domain of

    a st(x)

    b ts(x)

> **Talking point**
> Which values do you need to exclude from the domain of a fractional function?

12 p and q are functions such that
$p(x) = x^3 - 3x + 2$
and
$q(x) = \dfrac{1}{x}$

    a Show that $(x - 1)$ is a factor of $x^3 - 3x + 2$.

    b Factorise $x^3 - 3x + 2$.

    c Hence, find the domain of qp(x).

Chapter 5: Functions

5.3 Inverse functions

The inverse function reverses the effect of the original function.

$f^{-1}(x)$ is the inverse function of $f(x)$.

Example 1

Find the inverse function of $f(x) = 3x + 2$

$x \rightarrow \boxed{\times 3} \rightarrow \boxed{+2} \rightarrow 3x+2$

$\dfrac{x-2}{3} \leftarrow \boxed{\div 3} \leftarrow \boxed{-2} \leftarrow x$

The inverse function of $f(x) = 3x + 2$ is $f^{-1}(x) = \dfrac{x-2}{3}$

When the inverse function $f(x)$ is the same as the original function $f^{-1}(x)$, the function f is a **self-inverse function**.

When a function $f(x)$ is self-inverse, $f(f(x)) = x$

Example 2

Show that $f(x) = 3 - x$ is a self-inverse function.

$x \rightarrow \boxed{\times -1} \rightarrow \boxed{+3} \rightarrow 3-x$ — $-1(x-3) = -x+3 = 3-x$

$3-x \leftarrow \boxed{-1} \leftarrow \boxed{-3} \leftarrow x$

The inverse function of $f(x) = 3 - x$ is $f^{-1}(x) = 3 - x$

$f(f(x)) = f(3-x)$

$\qquad = 3 - (3-x)$

$\qquad = 3 - 3 + x$

$\qquad = x$

So $f(x) = 3 - x$ is self-inverse. — To show that $f(x)$ is self-inverse, show that $f(f(x)) = x$.

Practice

1 Find the inverse of each function.

   a $f(x) = 5x + 4$ **b** $g(x) = 2x - 1$ **c** $h(x) = \dfrac{x}{2} + 3$

2 Find the inverse of each function.

   a $f(x) = 3(x - 2)$ **b** $g(x) = \dfrac{x-5}{2}$ **c** $h(x) = \dfrac{3x-4}{5}$

Chapter 5: Functions

3   $f(x) = 5(x + 3)$ and $g(x) = \frac{x}{3} - 1$

    a   Find

        i $f^{-1}(10)$ ii $g^{-1}(2)$ iii $g^{-1}(-1)$

    b   Work out $f^{-1}(x) + g^{-1}(x)$.

4   Find the inverse of each function.

    a $f(x) = -x$ b $g(x) = 1 - x$ c $h(x) = x$

    What do these three functions have in common?

Hint for Q4c

$x \rightarrow +0 \rightarrow x$

5   **Reasoning**

    a   Find

        i 2^{-1} ii $\left(\frac{1}{2}\right)^{-1}$ iii n^{-1} iv $\left(\frac{1}{n}\right)^{-1}$

    b   What is the inverse of 'find the reciprocal'?

    c $h(x) = \frac{1}{x}$

    Show that $h(x)$ is self-inverse.

    d   Write down the domain of the function $h(x)$.

6   Find the inverse of each function.

    a $f(x) = \frac{2(x+1)}{3}$ b $g(x) = \frac{3(x+1)}{4}$ c $h(x) = \frac{2(3x+1)}{5}$

7   $f(x) = \frac{2(5x-1)}{3}$ and $g(x) = 4x + 7$

    Work out $f^{-1}(1) - g^{-1}(1)$

8   $f(x) = 2x - 3$

    a   Sketch the graphs of $f(x)$ and $f^{-1}(x)$ on the same axes.

    b   Label the coordinates of the intersection of the two graphs on the sketch.

    c   Write down the domain of $f(x)$ and of $f^{-1}(x)$.

Hint for Q8b

Solve $f(x) = f^{-1}(x)$

9   $f(x) = 4(x + 1)$

    Sketch the graphs of $f(x)$ and $f^{-1}(x)$ on the same axes. Label the point of intersection and the intercepts.

💬 **Talking point**

What do you notice about the x- and y-coordinates of the points of intersection of $f(x)$ and $f^{-1}(x)$?

10   $f(x) = 2x + 1$

    a   Draw graphs of $f(x)$ and $f^{-1}(x)$ on the same axes.

    b   Draw the line $y = x$ on the graph.

    c   What transformation maps $f(x)$ on to $f^{-1}(x)$?

Hint for Q10a

Draw an accurate graph on squared paper.

Chapter 5: Functions

d Reasoning Is the answer to part **c** also true for the functions in questions **8** and **9**?

Copy and complete these sentences.

The graph of f(x) is a _____ of f⁻¹(x) in the line _____.

The intersection of the graphs of f(x) and f⁻¹(x) lies on the line _____.

11 Reasoning

a Show that $x = \sqrt{4}$ has two solutions.

b Copy and complete this mapping diagram.

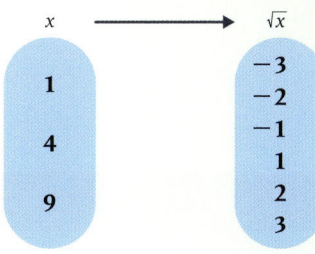

c Show that $x \to \sqrt{x}$ is not a function on the range $\{\sqrt{x} \in \mathbb{R}\}$.

d Draw a mapping diagram for $x \to \sqrt{x}$ on the range $\{\sqrt{x} \in \mathbb{R} : \sqrt{x} \geq 0\}$.

e Show that $x \to \sqrt{x}$ is a function in the range $\{\sqrt{x} \in \mathbb{R} : \sqrt{x} \geq 0\}$.

12 Problem-solving Find the inverse of each function. Write down any restrictions on the domain and range of the inverse function.

a $f(x) = x^2$ **b** $g(x) = x^3$ **c** $h(x) = x^4$

13 Problem-solving Sketch the graph of $f(x) = x^2$ and $f^{-1}(x)$ on the domain $\{x \in \mathbb{R} : x \geq 0\}$ and range $\{f^{-1}(x) \in \mathbb{R} : f^{-1}(x) \geq 0\}$.

Hint for Q13
Use the result from **Q10d**.

14 Find the inverse of each function.

a $f(x) = \sqrt{x}$ **b** $g(x) = 3 + \sqrt{x}$ **c** $h(x) = 5 + \sqrt{x+1}$

Hint for Q14b

$x \to \boxed{\sqrt{}} \to \boxed{+3} \to \sqrt{x} + 3$

$\boxed{^2} \leftarrow \boxed{-3} \leftarrow x$

15 Reasoning $f(x) = (4-x)^2 - 2$ and $g(x) = 4 - \sqrt{x+2}$

a Find fg(x)

b What does the answer to part **a** imply about the functions f(x) and g(x)?

Exam-style question

16 Given that
$f(x) = 2 + \sqrt{x+1}$
sketch the graph of $y = f^{-1}(x)$.

Clearly mark the exact coordinates of the turning point and the exact coordinates of any intersections with the coordinate axes. **(5 marks)**

Chapter 5: Functions

5.4 Transforming functions

Chapter 4 covers the effect on a graph when transforming a function. Transforming a function also has an effect on its equation.

To find the equation for

- $f(x) + a$, add a to the whole function
- $f(x) - a$, subtract a from the whole function
- $af(x)$, multiply the whole function by a.

To find the equation for

- $f(x + a)$, substitute $(x + a)$ for x in $f(x)$
- $f(x - a)$, substitute $(x - a)$ for x in $f(x)$
- $f(ax)$, substitute ax for x in $f(x)$.

Example

$f(x) = x^2(x + 1)$

Write the algebraic equation for

a $y = f(x) - 3$ **b** $y = f(x - 3)$ **c** $y = f(3x)$ **d** $y = -3f(x)$

a $f(x) = x^2(x + 1)$

Subtract 3 from $f(x)$. → $f(x) - 3 = x^2(x + 1) - 3$

Expand the brackets. → $y = x^3 + x^2 - 3$

b $f(x) = x^2(x + 1)$

Substitute $(x - 3)$ for x in $f(x)$. → $f(x - 3) = (x - 3)^2((x - 3) + 1)$

Simplify. → $y = (x - 3)^2(x - 2)$

$= x^3 - 8x^2 + 21x - 18$

c $f(x) = x^2(x + 1)$

Substitute $3x$ for x in $f(x)$. → $f(3x) = (3x)^2(3x + 1)$

$y = 9x^2(3x + 1)$

$= 27x^3 + 9x^2$

d $f(x) = x^2(x + 1)$

Multiply $f(x)$ by -3. → $-3f(x) = -3x^2(x + 1)$

$y = -3x^2(x + 1)$

$= -3x^3 - 3x^2$

Chapter 5: Functions

Practice

1 $f(x) = x^2$

Write the algebraic equation for

  a $y = f(x) - 2$ **b** $y = f(x - 2)$ **c** $y = f(2x)$ **d** $y = -2f(x)$

2 $f(x) = 3x - 2$

Write the algebraic equation for

  a $y = f(x) + 1$ **b** $y = f(x + 2)$ **c** $y = f(-2x)$ **d** $y = 2f(x)$

3 $f(x) = x^2 + 5$

Write the algebraic equation for

  a $y = f(x) - 3$ **b** $y = f(x + 1)$ **c** $y = f(3x)$ **d** $y = -2f(x)$

4 $f(x) = x(x - 2)$

Write the algebraic equation for

  a $y = f(x) + 2$ **b** $y = f(x - 2)$ **c** $y = f(-2x)$ **d** $y = 4f(x)$

5 $f(x) = x(x + 1)(x - 1)$

Write the algebraic equation for

  a $y = f(x) + 1$ **b** $y = f(x - 1)$ **c** $y = f(-2x)$ **d** $y = -2f(x)$

6 $f(x) = 2x + 1$

Write the algebraic equation for

  a $y = f(2x) - 1$ **b** $y = 3f(x + 2)$ **c** $y = f(-2x + 3)$

7 $f(x) = x^2$

Write the algebraic equation for

  a $y = f(3x) + 1$ **b** $y = 2f(x - 1)$ **c** $y = f(2x + 1)$

8 $f(x) = x^2 - x$

Write the algebraic equation for

  a $y = 2f(x + 1)$ **b** $y = f(3x - 1)$ **c** $y = f(2x) + 1$

> **Talking point**
>
> How do you know when to multiply, add to or subtract from the whole function?
>
> How do you know when to substitute for x in the function?

9 $f(x) = 4^x$

Write the algebraic equation for

a $y = f(2x - 3)$ **b** $y = f(2x) + 3$ **c** $y = 2f(x + 3)$

10 $f(x) = \dfrac{2}{x^2}$

Write the algebraic equation for

a $y = f(2x) - 1$ **b** $y = 2f(x - 1)$ **c** $y = f(-2x + 1)$

11 Graph G is a translation of $y = f(x)$.

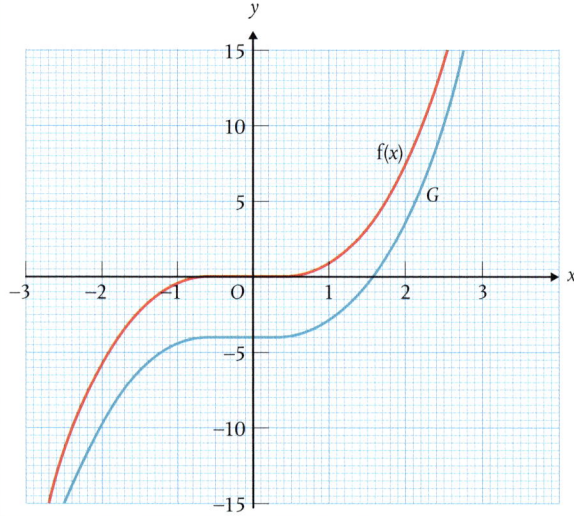

a Write down the equation of $y = f(x)$.

b Write down the equation of graph G.

Chapter 5: Functions

Exam-style question

12 The grid shows graph A and graph B.

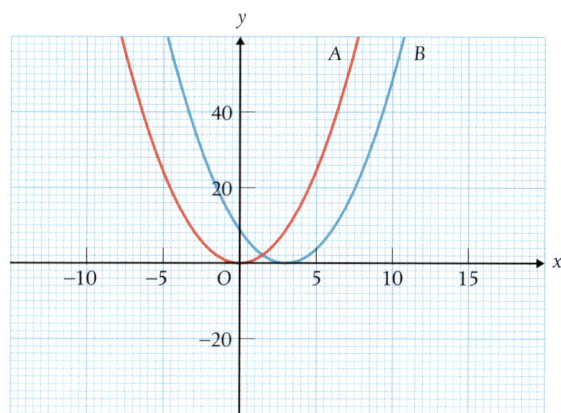

a Write down an equation for graph A. **(1 mark)**

Graph B is a translation of graph A.

b Write down an equation for the graph B. **(2 marks)**

13 Graph A is a transformation of $y = f(x)$.

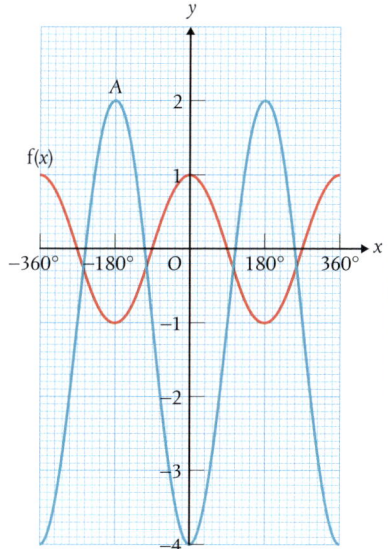

a Write down the equation of $y = f(x)$.

b Write down the equation of graph A.

14 Reasoning Here are some graphs.

Graph A
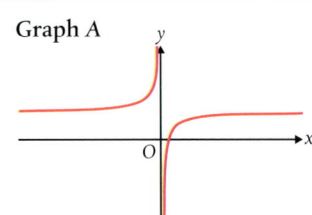

Graph B
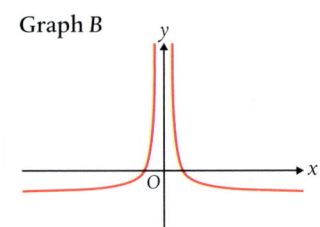

Graph C
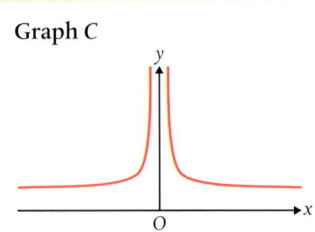

Graph D
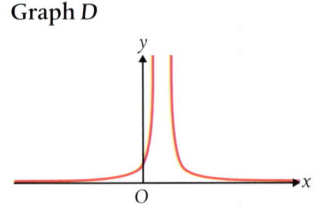

Graph E
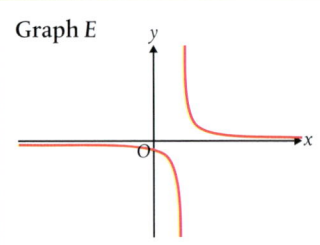

Graph F
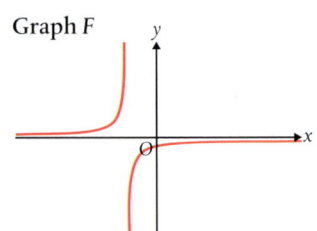

Here are the equations of the graphs.

i $y = \frac{1}{x^2} + 1$ ii $y = \frac{1}{x-2}$ iii $y = -\frac{1}{x} + 2$

iv $y = -\frac{1}{x+2}$ v $y = \frac{1}{(x-1)^2}$ vi $y = \frac{1}{x^2} - 1$

Match each graph to its equation.

15 $f(x) = \frac{1}{x^2}$

The graph of $y = f(x)$ is translated so the asymptotes are at $x = 2$ and $y = -1$.

Write down the equation for the transformed function in the form $y = \frac{1}{(x+a)^2} + b$

Hint for Q15

Work out the vector of the translation.

Maths challenge

Find the next number in this sequence: 13, 28, 520, 133, 738, 882, 1032, 36, ?

Katherine Johnson worked on several NASA space missions and used advanced algebraic techniques to work out the trajectories of rocket flights, including parts of the Apollo mission to the Moon. When she started, calculations had to be done by hand and she led a team of women, including many African Americans, known as 'computers'. Later she pioneered the use of digital computers in this field.

Chapter 6:
Equations and inequalities

In this chapter you will:
- solve equations involving algebraic fractions
- solve trigonometric equations
- solve equations given in function notation
- solve quadratic equations
- solve quadratic equations involving trigonometric functions
- solve simultaneous equations by substitution
- solve simultaneous equations where one equation is quadratic
- solve linear and quadratic inequalities in a single variable
- solve problems using linear programming

Prior knowledge
- add, subtract, multiply and divide algebraic fractions
- solve simultaneous equations by elimination
- use function notation
- find inverse functions and composite functions
- sketch sine, cosine and tangent graphs
- know exact values of trigonometric functions
- solve quadratic equations by factorising, using the quadratic formula and completing the square
- sketch graphs of quadratic functions
- solve linear simultaneous equations algebraically
- know the equation of a circle centre (a, b) radius r
- use set notation

Chapter 6: Equations and inequalities

6.1 Solve equations

An algebraic fraction has an unknown in either the numerator, the denominator or both.

To solve equations with algebraic fractions, first write one or both sides as fractions in their simplest form.

Example 1

Solve

$$\frac{3x-5}{2} + \frac{4x}{3} = 6$$

- Write both fractions over a common denominator.

$$\frac{3(3x-5)}{6} + \frac{2 \times 4x}{6} = 6$$

- Simplify the numerators.

$$\frac{9x-15}{6} + \frac{8x}{6} = 6$$

- Add the fractions.

$$\frac{9x-15+8x}{6} = 6$$

$$\frac{17x-15}{6} = 6$$

- Multiply both sides by the denominator.

$$17x - 15 = 36$$
$$17x = 51$$
$$x = 3$$

Trigonometric equations may have more than one solution. Use a graph to find them all.

Example 2

Solve

$$3 \sin x - 2 = 0$$

for $-180° \leq x \leq 180°$

$$3 \sin x - 2 = 0$$
$$3 \sin x = 2$$

- Rearrange to make sin x the subject.

$$\sin x = \frac{2}{3}$$

- Solve for between 0° and 90°.

$$x = \sin^{-1}\left(\frac{2}{3}\right)$$
$$= 41.8 \text{ (1 d.p.)}$$

- Sketch the graph of $y = \sin x$ for $-180° \leq x \leq 180°$
  Draw the line $y = \frac{2}{3}$.

- Mark on the angle 41.8°.

- Use the symmetry of the graph to find other solutions.

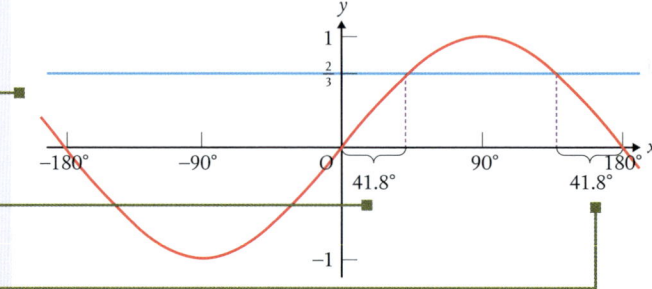

$$180° - 41.8° = 138.2°$$

Solutions are $x = 41.8°$ and $x = 138.2°$ (1 d.p.)

Chapter 6: Equations and inequalities

Practice

1 Solve

   a $\dfrac{2x+1}{3} + \dfrac{5x}{4} = 8$ **b** $\dfrac{3x-2}{4} - \dfrac{x}{3} = 2$

   c $\dfrac{4x+1}{7} + \dfrac{x-3}{2} = 4$ **d** $\dfrac{2x+7}{2} + \dfrac{3-4x}{5} = \dfrac{7}{2}$

2 Solve

   a $\dfrac{5x+2}{4} - \dfrac{2x-1}{3} = \dfrac{x+10}{6}$ **b** $\dfrac{2x+3}{4} - \dfrac{4x-2}{3} = \dfrac{x+18}{4}$

> **Hint for Q2**
> Write all the fractions with a common denominator. Multiply both sides by the common denominator.

3 Solve

   a $\dfrac{4}{x} + \dfrac{5}{x} = 7$ **b** $\dfrac{4}{x-3} - \dfrac{6}{x-3} = 5$

   c $\dfrac{5}{x} + \dfrac{3}{2x} = 8$ **d** $\dfrac{x-1}{2x} - \dfrac{3}{x} = 4$

> **Hint for Q3**
> Give any fraction or mixed number answers in their lowest terms.

4 The total resistance of a set of resistors in a parallel circuit is given by the formula
$\dfrac{1}{R} = \dfrac{1}{R_1} + \dfrac{1}{R_2}$

   a Find R_2 when $R_1 = 6$ and $R = 5$.

   b Find R when $R_1 = \dfrac{3}{4}$ and $R_2 = 5$.

   c **Reasoning** Find R_1 and R_2 when $R = 12$ and R_1 and R_2 are in the ratio $1:3$.

5 **Reasoning** The relationship between the focal length f of a lens, the image distance v and the object distance u is given by the formula
$\dfrac{1}{f} = \dfrac{1}{v} - \dfrac{1}{u}$

   a Write a formula for f when $u = 5v$ **b** Find u and v when $f = 15$ and $v = \dfrac{1}{2}u$.

6 Solve

   a $\dfrac{x^2 - 9}{x^2 - x - 6} = \dfrac{11}{10}$ **b** $\dfrac{x^2 + 3x + 2}{x^2 - 3x - 10} = \dfrac{1}{4}$

   c $\dfrac{3(x^2 - 1)}{x - 4} \times \dfrac{2}{x+1} = 24$ **d** $\dfrac{x^2 - 4x - 5}{x^2 - 1} \div \dfrac{x^2 - 3x - 10}{x+2} = 0.5$

> **Hint for Q6**
> Factorise and simplify first.

7 ▦ Solve

   $4\sin x - 1 = 0$

for $-180° \leq x \leq 180°$.

Give the solutions to one decimal place.

8 **Problem-solving** Solve

   $2\sin x - 1 = 0$

for $-180° \leq x \leq 180°$.

9 **Problem-solving** Solve

   $2\cos x - 1 = 0$

for $-180° \leq x \leq 180°$.

Chapter 6: Equations and inequalities

10 **Problem-solving** Solve

$$2\tan\theta - 7 = 0$$

for $0° \leq \theta \leq 360°$.

Exam-style question

11 Solve $\sqrt{2}\cos x° - 1 = 0$ for $-180° \leq x \leq 180°$. **(2 marks)**

12 **a** Draw the graph of $y = \cos x°$ for $0° \leq x \leq 360°$.

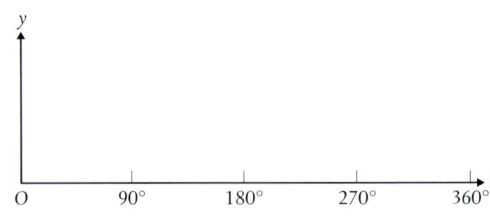

(2 marks)

b Diane says, '$3\cos x° - 2 = 2.5$ has no solutions'.

Diane is correct.

Explain why. **(1 mark)**

Talking point

For what values of k does $\cos x = k$ have solutions?

For what values of k does $\sin x = k$ have solutions?

13 **a** **Reasoning** Which of these trigonometric equations have no solutions for $-180° \leq x \leq 180°$?

    i $2\sin x - 4 = 1$ **ii** $3\tan x - 5 = 1$

    iii $4\cos x - 2 = 2$ **iv** $4\sin x = -2$

    v $2\cos x = \sqrt{3}$ **vi** $\sqrt{3}\cos x = 2$

b Solve the equations that have solutions.

14 **Reasoning** The table shows some values of x and y that satisfy the equation

$$y = a\cos x + b$$

x	0°	60°	90°
y	7	6	5

a Find the value of y when $x = 30°$.   **b** Show that $y = 0$ has no solutions.

15 **Problem-solving**

a $f(x) = 7x + 3$. Solve $f^{-1}(a) = 1$.

b $g(x) = \dfrac{x-1}{4}$. Solve $g^{-1}(a) = 3$.

Hint for Q16

$fg(x)$ is $f(g(x))$

16 $f(x) = 2x + 1$, $g(x) = \cos x$

Solve $fg(a) = -\dfrac{1}{2}$ for $0° \leq x \leq 360°$. Give the answers to the nearest degree.

17 **Problem-solving** $f(x) = 3(x - 2)$, $g(x) = 2x + 3$

Given that $f^{-1}(a) = g^{-1}(a)$, work out the value of a.

Chapter 6: Equations and inequalities

6.2 Solve quadratic equations

Equations that involve algebraic fractions – and some real-life situations – may lead to solving quadratic equations.

The **roots** of a quadratic function f(x) are the solutions to f(x) = 0.

Solve quadratic equations by factorising, using the quadratic formula or completing the square.

Example 1

Find the roots of the function

$$f(x) = 3x^2 + 2x - 6$$

Give the answers in surd form.

$$3x^2 + 2x - 6 = 0$$

$$x = \frac{-b \pm \sqrt{b^2 - 4ac}}{2a}$$

$$= \frac{-2 \pm \sqrt{2^2 - 4 \times 3 \times -6}}{2 \times 3}$$

$$= \frac{-2 \pm \sqrt{4 + 72}}{6}$$

$$= \frac{-2 \pm \sqrt{76}}{6}$$

$$= \frac{-2 \pm 2\sqrt{19}}{6}$$

$$= \frac{-1 \pm \sqrt{19}}{3}$$

The roots are $x = \frac{-1 + \sqrt{19}}{3}$ and $x = \frac{-1 - \sqrt{19}}{3}$.

'Answers in surd form' is a clue to use the quadratic formula or completing the square.

$\sqrt{76} = \sqrt{4}\sqrt{19} = 2\sqrt{19}$

Example 2

Solve

$$9x^2 + 4x - 3 = 0$$

by completing the square.

$$9\left(x^2 + \frac{4}{9}x\right) - 3 = 0$$

$$9\left[\left(x + \frac{2}{9}\right)^2 - \frac{4}{81}\right] - 3 = 0$$

$$9\left(x + \frac{2}{9}\right)^2 - \frac{4}{9} - 3 = 0$$

$$9\left(x + \frac{2}{9}\right)^2 = \frac{31}{9}$$

$$\left(x + \frac{2}{9}\right)^2 = \frac{31}{81}$$

$$x + \frac{2}{9} = \pm \frac{\sqrt{31}}{9}$$

$$x = \frac{-2 - \sqrt{31}}{9} \text{ and } x = \frac{-2 + \sqrt{31}}{9}$$

Write as $a\left(x^2 + \frac{b}{a}x\right) + c$

Complete the square.

Divide both sides by 9.

Square root both sides.

Chapter 6: Equations and inequalities

Example 3

Solve

$$\frac{x+2}{x-1} + \frac{x}{x+4} = 3$$

$$\frac{(x+4)(x+2)}{(x+4)(x-1)} + \frac{x(x-1)}{(x+4)(x-1)} = 3$$

Add the fractions.

$$\frac{(x+4)(x+2) + x(x-1)}{(x+4)(x-1)} = 3$$

Expand the brackets.

$$\frac{x^2 + 6x + 8 + x^2 - x}{x^2 + 3x - 4} = 3$$

Multiply both sides by $x^2 + 3x - 4$.

$$x^2 + 6x + 8 + x^2 - x = 3(x^2 + 3x - 4)$$

$$x^2 + 6x + 8 + x^2 - x = 3x^2 + 9x - 12$$

$$x^2 + 4x - 20 = 0$$

Complete the square.

$$[(x+2)^2 - 4] - 20 = 0$$

$$(x+2)^2 = 24$$

$$x + 2 = \pm\sqrt{24}$$

$$x + 2 = \pm\sqrt{4}\sqrt{6}$$

$$x + 2 = \pm 2\sqrt{6}$$

$$x = -2 - 2\sqrt{6} \text{ and } x = -2 + 2\sqrt{6}$$

Practice

1 Find the roots of each function. Give the answers in surd form.

   a $f(x) = 2x^2 + 5x - 4$ b $f(x) = 5x^2 - 4x - 2$ c $f(x) = 2x^2 + 3x - 1$

2 Solve by completing the square.

   a $3x^2 + 7x - 6 = 0$ b $4x^2 + 4x - 1 = 0$ c $6x^2 - 3x - 4 = 0$

Hint for Q3
Represent one of the odd numbers as $2n - 1$.

3 **Reasoning** The product of two consecutive odd numbers is 3363.
   Find the two numbers.

4 **Problem-solving** The sum of two consecutive square numbers is 1513.
   Find the two numbers.

Hint for Q5
Draw a diagram.

5 **Problem-solving** A right-angled triangle has hypotenuse 12 cm.
   One of the shorter sides is 4 cm longer than the other.
   Find the lengths of the two shorter sides, to the nearest millimetre.

💬 **Talking point**
Which solution to the quadratic equation did you ignore, and why?

Chapter 6: Equations and inequalities

6 **Problem-solving** This triangle has area 35 m².

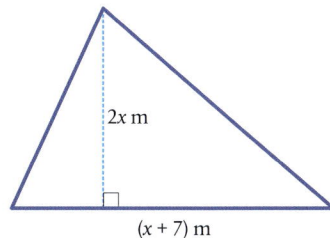

Show that the height of the triangle is equal to $-7 + 3\sqrt{21}$ metres.

7 **Reasoning** The diagram shows the cable of a suspension bridge.

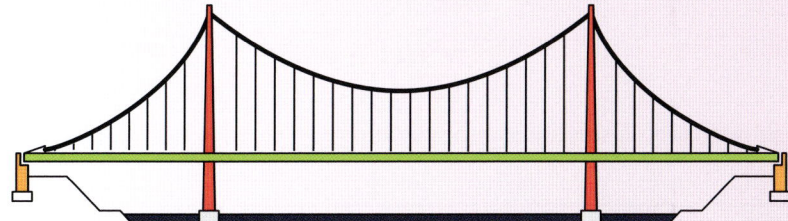

The curve of the cable in the middle span is modelled by the function $h(x) = 0.01x^2 + 3$, where x is the distance from the centre of the bridge and $h(x)$ is the height of the cable above the road at point x.

a   What does the value 3 in the function $h(x)$ represent?

b   The towers at each end of the bridge are 19 m tall.

Find the length of the bridge in metres.

8 **Problem-solving** The sketch graph shows the trajectory (path) of a golf ball in metres.

$g(x) = -(0.05x - 3)^2 + 9$

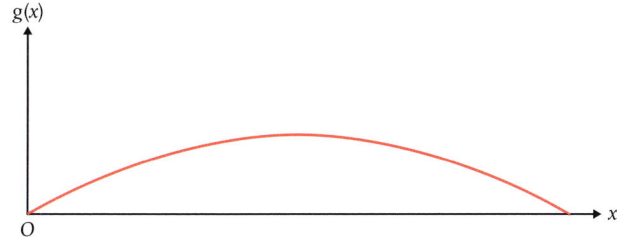

a   Work out how far from the starting point the ball hits the ground.

b   Find the maximum height the ball reaches.

9 $f(x) = 2x - 3$ and $g(x) = x^2 + 1$

a   Work out

   i $fg(x)$ ii $gf(x)$

b   Solve $fg(x) = gf(x)$.

Chapter 6: Equations and inequalities

10 Solve

a $\dfrac{6}{x} = \dfrac{4x+1}{3}$

b $\dfrac{7}{2x} = \dfrac{3x-11}{4}$

c $\dfrac{6}{x+4} + \dfrac{4}{x+2} = 2$

d $\dfrac{1}{x+1} + \dfrac{x-2}{3} = 4$

11 a **Reasoning** Show that the equation $\dfrac{5}{x-2} + \dfrac{3x+1}{2x} = 2$ can be rearranged to give $x^2 - 13x + 2 = 0$.

b Hence, solve the equation $\dfrac{5}{x-2} + \dfrac{3x+1}{2x} = 2$.

12 a **Reasoning** Show that the equation $\dfrac{5}{2x+1} + \dfrac{3}{x-2} = 1$ can be rearranged to give $2x^2 - 14x + 5 = 0$.

b Hence, show that both solutions to the equation $\dfrac{5}{2x+1} + \dfrac{3}{x-2} = 1$ satisfy $0 < x < 7$.

Exam-style question

13 a Write $\dfrac{x-2}{2} = \dfrac{4}{x-3}$ in the form $ax^2 + bx + c = 0$. **(2 marks)**

b Hence, or otherwise, solve $\dfrac{x-2}{2} = \dfrac{4}{x-3}$.

Give the answer in the form $\dfrac{p \pm \sqrt{q}}{2}$, where p and q are integers. **(2 marks)**

14 a Substitute $u = x^2$ into the quartic equation $x^4 - 7x^2 + 12 = 0$.

b Solve the equation from part **a** to find two values of u.

c **Reasoning** Hence, solve the equation $x^4 - 7x^2 + 12 = 0$ for x.

Hint for Q14c
A quartic equation has up to four roots.

15 Solve the equation $x - \sqrt{x} = 6$ by substituting $v = \sqrt{x}$.

16 Solve the equation $2\sin^2 x + \sin x = 1$ for $-180° \le x \le 180°$ by substituting $u = \sin x$.

Hint for Q16
$\sin^2 x = (\sin x)^2$

Exam-style question

17 a Solve $2x^2 = x + 1$ **(3 marks)**

b Given that $\sin^2 x + \cos^2 x = 1$,
solve $1 - 2\sin^2 x - \cos x = 0$ for $0° \le x < 360°$ **(4 marks)**

Chapter 6: Equations and inequalities

6.3 Solve simultaneous equations

Solve simultaneous equations by:

- elimination – make the coefficients of one variable the same in both equations, and then either add or subtract the equations to eliminate this variable; for example, add these simultaneous equations to eliminate the y term:

  $x + 2y = 25$
  $3x - 2y = 10$

- substitution – substitute an expression for x or y from one equation into the other equation.

A pair of quadratic and linear simultaneous equations can have two solutions, one solution or no solution.

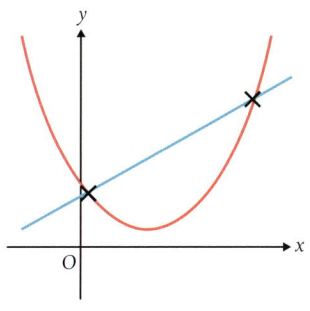
Two solutions

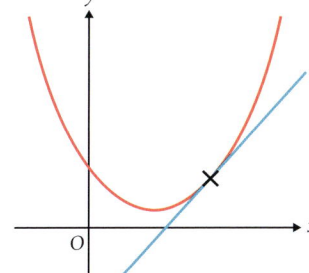
One solution

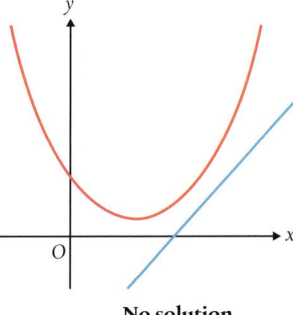
No solution

Example 1

Solve these simultaneous equations by substitution.

$4x + y = 8$

$6x - 2y = 1$

Give the solutions to two decimal places.

$4x + y = 8$
$\quad y = 8 - 4x$ — Rearrange one of the equations to make y the subject.

$\quad 6x - 2y = 1$
$6x - 2(8 - 4x) = 1$ — Substitute the expression for y into the other equation.

$6x - 16 + 8x = 1$ — Expand and simplify.

$\quad 14x = 17$

$\quad x = \dfrac{17}{14}$ — Substitute this x-value into the expression for y.

$\quad y = 8 - 4x$

$\quad y = 8 - 4 \times \dfrac{17}{14}$

$\quad y = 8 - \dfrac{34}{7}$

$\quad = \dfrac{56 - 34}{7}$

$\quad = \dfrac{22}{7}$ — Only round the two solutions to 2 d.p. at the end, to keep them as accurate as possible.

$x = 1.21, y = 3.14$ (2 d.p.)

Chapter 6: Equations and inequalities

Example 2

The sum of two numbers is 19 and the sum of their squares is 241.

Find the two numbers.

Let the two numbers be x and y.

Write equations to represent the information in the question.
$$x + y = 19$$
$$x^2 + y^2 = 241$$

$$x = 19 - y$$

Solve the simultaneous equations by substitution.
$$(19 - y)^2 + y^2 = 241$$

Expand and simplify.
$$361 - 38y + y^2 + y^2 = 241$$

Solve the quadratic.
$$2y^2 - 38y + 120 = 0$$
$$y^2 - 19y + 60 = 0$$
$$(y - 4)(y - 15) = 0$$
$$y = 4 \text{ or } y = 15$$

Substitute each y-value into the expression for x.
When $y = 4$, $x = 19 - y = 15$
When $y = 15$, $x = 19 - y = 4$

Give the answer in the context of the question.
The two numbers are 4 and 15.

 Talking point

Which substitution did you use? Which other substitution could you have used?

 Talking point

Could you also have solved these by elimination?

Practice

1 Solve each pair of linear simultaneous equations by substitution. Give any non-integer solutions to 2 decimal places.

a $2x + y = 8$
$3x - 2y = 5$

b $5(x + 3) - y = 6$
$4x - 3y = -16$

c $4x - 3y = 11$
$4y = 3x + 2$

Chapter 6: Equations and inequalities

2 **Reasoning** Sam swims 2.3 km in 1 hour.

Some of this swim is front crawl at 3.2 km/h and the rest is breaststroke at 2 km/h.

a Let distance swum in front crawl be x and distance swum in breaststroke be y.

Write an equation connecting x and y.

b Using the formula time = $\frac{\text{distance}}{\text{speed}}$, show that

  i the time swimming front crawl is $\frac{x}{3.2}$.

  ii the time swimming breaststroke is $\frac{y}{2}$.

c Hence write another equation connecting x and y.

d What distance did Sam swim using breaststroke?

3 **Reasoning** Mel thinks of a fraction.

When Mel adds 1 to both numerator and denominator the answer is $\frac{3}{5}$.

When Mel subtracts 4 from both numerator and denominator the answer is $\frac{1}{5}$.

By writing and solving two simultaneous equations, find the fraction Mel thinks of.

Hint for Q3
Let the fraction be $\frac{x}{y}$.

4 **Problem-solving** The function g is such that

$$g(x) = ax + b$$

where a and b are constants.

$g(3) = 1$ and $g^{-1}(3) = 4$

Find the value of a and the value of b.

5 Solve each pair of simultaneous equations by substitution.
Give any non-integer solutions as fractions.

a $y - x^2 = 3$

  $2x + 4y = 12$

b $y - x^2 = 1$

  $y = 2x + 1$

c $y = 2x^2 + 3$

  $y = x + 4$

Hint for Q5
Write the equations in the form $y = \ldots$

6 The diagram shows a triangle drawn inside a rectangle.

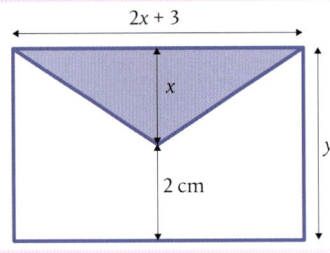

The length of the rectangle is $2x + 3$.

The width of the rectangle is y.

The area of the triangle is $6\,cm^2$.

The height of the triangle is 2 cm less than y.

a **Reasoning** Show that this problem can be represented by the simultaneous equations

$$2x^2 + 3x = 12$$

$$y = x + 2$$

b Solve the equations from part **a** to find the width of the rectangle y to the nearest millimetre.

7 **Problem-solving** The sketch graph shows two lines, L_1 and L_2

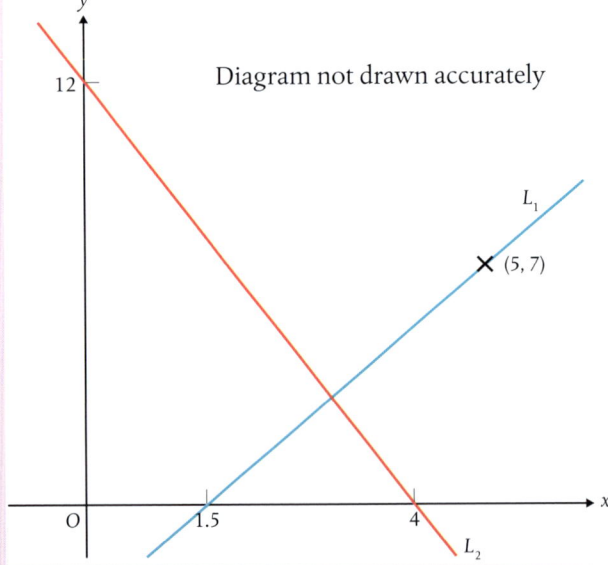

Find the coordinates where lines L_1 and L_2 intersect.

Chapter 6: Equations and inequalities

8 Solve each pair of simultaneous equations by substitution.
Give any non-integer solutions as fractions.

a $y + \frac{3}{x} = 6$

   $y = 3x$

b $y + \frac{3}{x-2} = 1$

   $y = 4x$

c $\frac{y}{2} + \frac{1}{x+2} = 6$

   $y + 3x = 1$

9 **Problem-solving** Find the two points of intersection of the graphs

   $y - \frac{2}{x+1} = 5$

   $y - x = 6$

   Give the coordinates in surd form.

10 **Reasoning**

a Solve these simultaneous equations.

   $x^2 + y^2 = 25$

   $x + 2y = 2$

b Sketch graphs of the two equations on the same axes, and write the coordinates of their points of intersection.

11 By writing and solving two simultaneous equations, find the coordinates of the points of intersection of these two graphs.

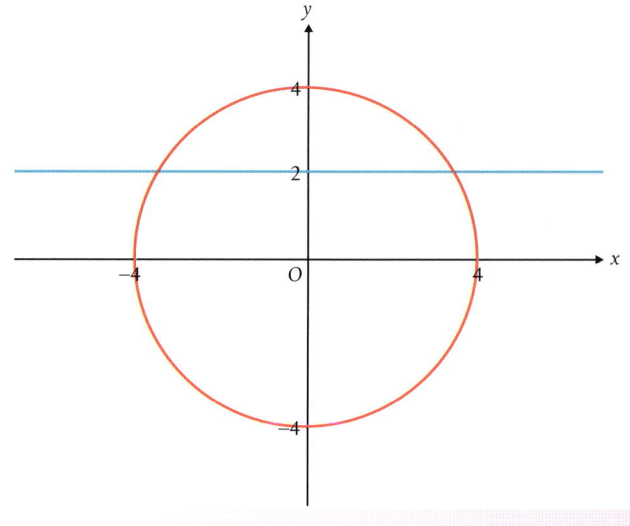

Chapter 6: Equations and inequalities

12 **Problem-solving** Find the coordinates of the points of intersection of the circle centre (0,0) radius $\sqrt{13}$ and the line $y = x + 1$.

13 **Problem-solving** Find the coordinates of the points of intersection of the circle centre (0,0) radius 3 and the line $y + 2x = 5$.

14 **Reasoning**

    a Show that the circle $x^2 + y^2 = 10$ and the line $y = 3x + 10$ intersect at exactly one point.

    b What does the result to part **a** imply about the relationship between the line $y = 3x + 10$ and the circle?

15 By sketching a graph, or otherwise, show that the simultaneous equations

$$x^2 + y^2 = 14$$

$$x + y = 6$$

have no solutions.

16 **Problem-solving** Find the coordinates of the points of intersection of the circle centre (0, 3) radius 2 and the line $x + y = 1$.

Exam-style question

17 Circle **C** has centre (4, 1) and radius 3.

Line **L** has equation $y = 2x - 8$.

Find the coordinates of the points of intersection between **C** and **L**.

Give the coordinates correct to 3 significant figures. **(7 marks)**

Chapter 6: Equations and inequalities

6.4 Solve inequalities

It is helpful to represent inequalities on a graph when solving problems.

Writing and drawing inequalities to represent a real-life problem is called **linear programming**.

The **feasible region** of the graph is the region of points that satisfy all the inequalities.

Example 1

Shona makes mugs to sell and Shauntae makes bowls to sell.

Shona can make up to 40 mugs a week.

Shauntae can make up to 30 bowls a week.

Let x be the number of mugs and let y be the number of bowls made in one week.

a Write down two inequalities to represent this information.

b In any one week this inequality is also true: $x + y > 45$

Describe, in context, what this inequality represents.

c By drawing suitable lines on a graph, show the feasible region for the values of x and y. Label the feasible region R.

a $0 \leq x \leq 40$

$0 \leq y \leq 30$

Up to 40 mugs, up to 30 bowls. The number of each has to be ≥ 0.

b $x + y > 45$ means that they make more than a total of 45 items each week.

c

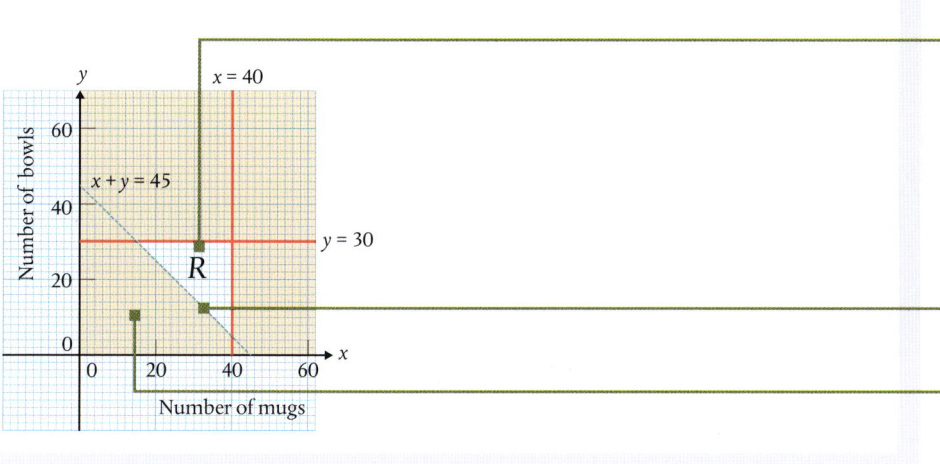

All the points in the feasible region, R, satisfy all the inequalities.

$x + y > 45$, so 45 is not included. Draw a dashed line.

*Shade the side of each line that **does not** satisfy each inequality.*

Chapter 6: Equations and inequalities

Use linear programming to work out how Shauntae and Shona in Example 1 can maximise their profit.

They make a profit of £5 on each mug and £7 on each bowl.

Their profit is given by the formula

$P = 5 \times$ number of mugs $+ 7 \times$ number of bowls

$P = 5x + 7y$

In linear programming, an **objective function** is the formula for the amount you want to maximise (or minimise).

To find the values of x and y that maximise the profit P, test the x- and y-values at the vertices of the feasible region to find which pairs give the greatest value of P.

Example 2

Using the objective function $P = 5x + 7y$ and the feasible region from the graph in Example 1, work out the maximum profit they can make in one week.

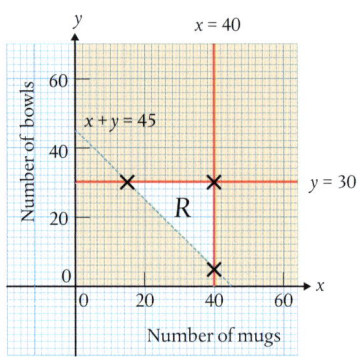

The vertices of the feasible region are (15, 30) (40, 5) and (40, 30)

(15, 30) → $P = 5x + 7y = 5 \times 15 + 7 \times 30 = £285$

(40, 5) → $P = 5x + 7y = 5 \times 40 + 7 \times 5 = £235$

(40, 30) → $P = 5x + 7y = 5 \times 40 + 7 \times 30 = £410$

The maximum profit they can make in one week is £410, when they make 40 mugs and 30 bowls.

Test each pair of x- and y-values to find the maximum value of P.

Chapter 6: Equations and inequalities

Example 3

On a coordinate grid, show the region that satisfies the inequalities

$-x^2 + 5x + 14 \geq y$ and $y > x + 2$

$-x^2 + 5x + 14 \geq y$

$-x^2 + 5x + 14 = (-x + 7)(x + 2)$ — Factorise the left-hand side.

$(-x + 7)(x + 2) = 0$ — Put the left-hand side equal to zero to find the roots.

$\Rightarrow x = -2, x = 7$

y-intercept $= 14$

Turning point is at $x = \dfrac{-2 + 7}{2} = 2.5$. — By symmetry, the x-coordinate of the turning point is halfway between the two roots.

When $x = 2.5$, $y = -2.5^2 + 5 \times 2.5 + 14 = 20.25$

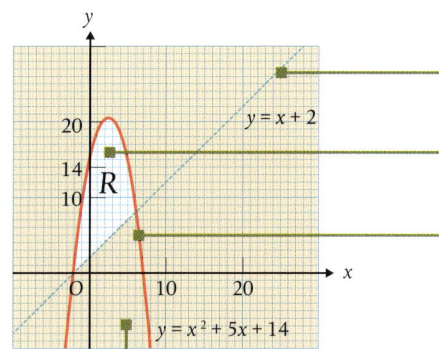

Label the region that satisfies **all** the inequalities R.

$y > x + 2$, so draw a dashed line.

Use the turning point, roots and y-intercept to draw the graph of the quadratic function.

Shade the side of each line/curve that **does not** satisfy each inequality.

Example 4

Solve the inequality

$2x^2 + x - 3 > 0$

$2x^2 + x - 3 > 0$

$2x^2 + x - 3 = (2x + 3)(x - 1)$ — Factorise the left-hand side.

$(2x + 3)(x - 1) = 0$ — Put the left-hand side equal to zero to find the roots.

$\Rightarrow x = -\dfrac{3}{2}, x = 1$

Chapter 6: Equations and inequalities

Sketch the graph of $y = 2x^2 + x - 3$, labelling the x-intercepts.

Highlight the x-values where $2x^2 + x - 3 > y$.

Circle any values not included. $2x^2 + x - 3 > 0$ does not include the values where $y = 0$

Write the set of x-values that satisfy the inequality using set notation.

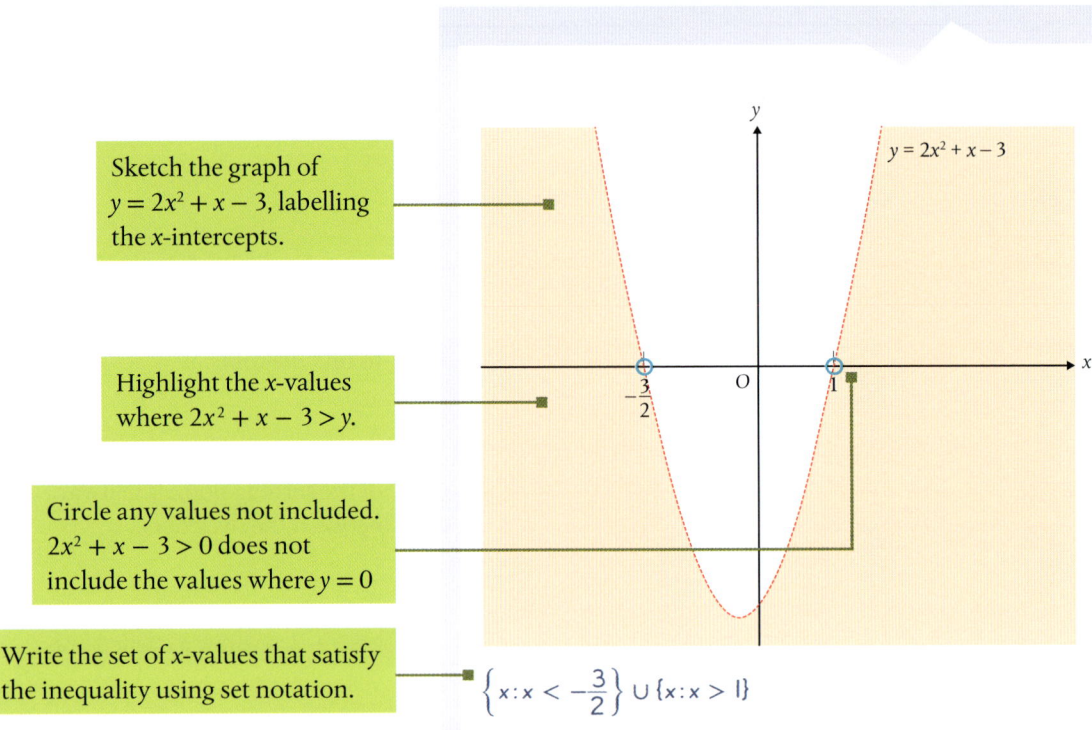

$\left\{x : x < -\dfrac{3}{2}\right\} \cup \{x : x > 1\}$

Practice

Hint for Q1

When multiplying or dividing both sides by a negative number, reverse the inequality sign.

1 Solve these inequalities

   a $-2x > 8$ b $-5x + 1 < -3$ c $-7 \leq -3x + 11$

2 **Reasoning** A farmer plants two crops, maize and wheat.
   She plants at least 30 hectares of maize.
   She plants at least twice as much wheat as maize.
   Let x be the number of hectares of maize.
   Let y be the number of hectares of wheat.

   a Write down two inequalities to represent this information.

      This inequality is also true

      $x + y \leq 140$

   b Describe, in context, what this inequality represents.

   c Draw a graph to show the feasible region for the values of x and y.

      Label the feasible region R.

3 The graph shows the feasible region, R, for a linear programming problem.

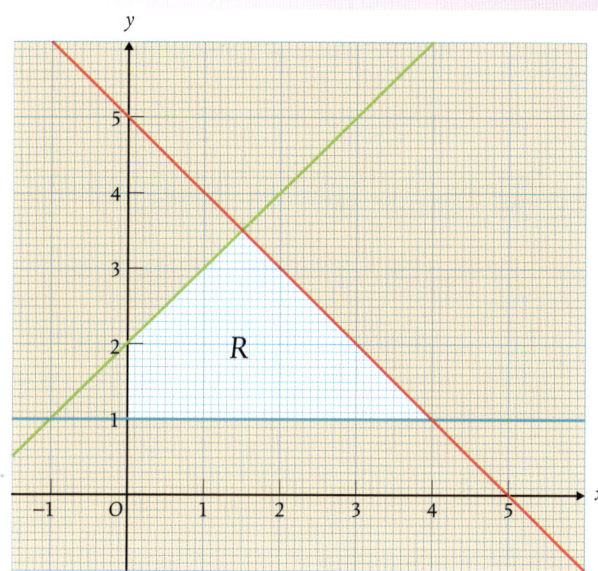

The objective function is $C = 5x + 8y$, where C is the cost.

a Find the values of x and y that minimise the objective function.

b Find the values of x and y that maximise the objective function.

4 **Problem-solving** Given that $x > 2$, $y \leq \frac{x}{3}$, $y > 0$ and $x - y \leq 7$, find the values of x and y that maximise the objective function $H = 4x + 5y$.

Exam-style question

5 A rental car company rents out two types of cars, petrol cars and electric cars.
They have 12 petrol cars.
They have 15 electric cars.
Let x be the number of petrol cars rented on one day.
Let y be the number of electric cars rented on one day.

a Write down two inequalities to represent this information. **(2 marks)**

On any one day the following inequalities are also true.
$x + y > 5$ and $x + y \leq 20$

b Describe, in context, what these inequalities represent. **(2 marks)**

c Use the grid below to show the feasible region for the values of *x* and *y*. Label the feasible region R.

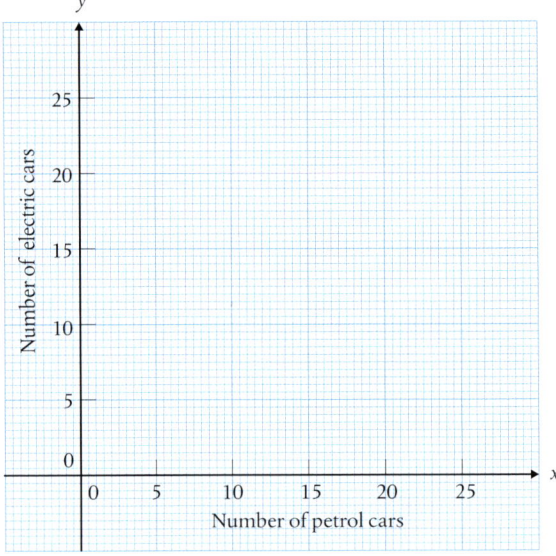

(4 marks)

d Why are there no negative values shown on this grid?
Give the answer in context. (1 mark)

The charge for renting a petrol car is £75 per day.
The charge for renting an electric car is £105 per day.

e Write down the objective function. (1 mark)

f Work out the maximum amount the company receives in charges on any one day. (3 marks)

6 Problem-solving A theatre sells adult tickets and child tickets.

For the theatre to make a profit, it must sell at least 70 adult tickets and 200 child tickets for a performance.

The theatre has 400 seats.

The price of an adult ticket is £32.

The price of a child ticket is £20.

Let *x* be the number of adult tickets sold.

Let *y* be the number of child tickets sold.

a Write the objective function in *x* and *y* for T, the amount of money made from selling tickets.

b How many of each ticket should be sold to maximise the value of T?

Chapter 6: Equations and inequalities

7   On a coordinate grid, show the region that satisfies the inequalities
    $x^2 - 3x - 10 \leq y$ and $y < x - 1$.

8   a   On a coordinate grid, show the x-values that satisfy the inequalities
        $-2x^2 + 3x + 20 \geq 0$ and $x + y \leq 2$.

    b   Write the set of values that satisfy the inequalities using set notation.

9   Here is a sketch graph of $y = 5x^2 + 7x + 2$.

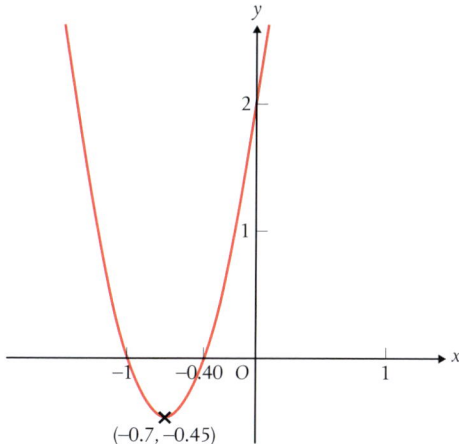

    Write the solution sets for x for each of these inequalities.

    a $5x^2 + 7x + 2 < 0$

    b $5x^2 + 7x + 2 \leq 0$

    c $5x^2 + 7x + 2 > 0$

    d $5x^2 + 7x + 2 \geq 0$

10 a   Draw the graph of $y = 4x^2 - 5x - 6$ on a graph paper grid.

    b   Find the values of x that satisfy $4x^2 - 5x - 6 \leq 0$.

    c   **Reasoning** By drawing a suitable line on the graph, find the values of x that satisfy $4x^2 - 5x - 3 \leq 0$.

11 Solve each inequality, giving the solution set for x.

    a $x^2 + x - 2 > 0$

    b $3x^2 - 13x + 4 \geq 0$

    c $-x^2 + 5x - 4 \geq 0$

Talking point
Why is it helpful to draw a graph to solve quadratic inequalities?

Hint for Q10c
Rearrange $4x^2 - 5x - 3 \leq 0$ into the form $4x^2 - 5x - 6 \leq a$, then draw the line $y = a$.

Chapter 6: Equations and inequalities

Exam-style question

12 a On the grid show, by shading, the region that satisfies all of these inequalities.

$2y + 4 \geq x$

$y \leq 6 - 3x$

$y \geq x^2 + 2x - 4$

Label the region R.

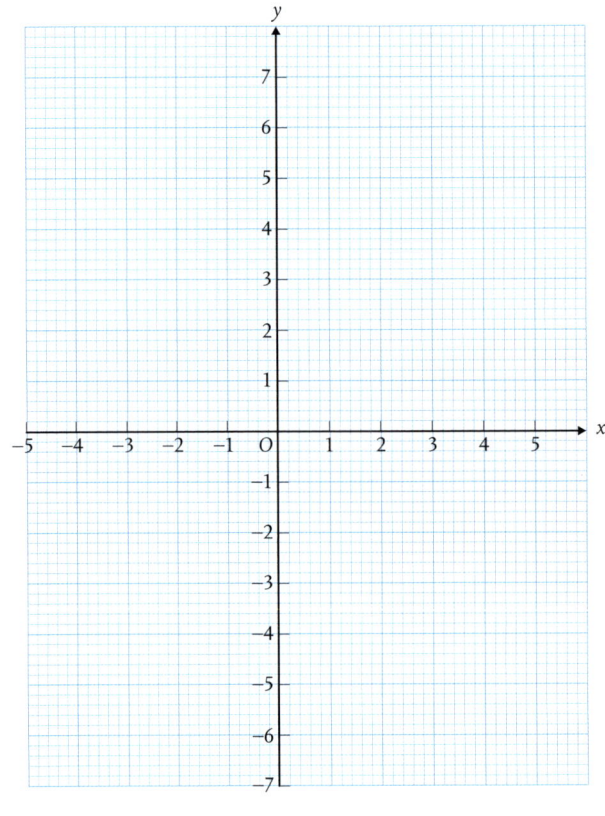

(4 marks)

b Hence, find the greatest value of x that satisfies all of these inequalities. (2 marks)

Maths challenge

Find the lowest positive integers for w, x, y and z.

w + x = y

9z = 4y

w - z = x

The ideas of trigonometry go back to ancient Egyptian and Babylonian mathematics. Neither of these cultures had a way of measuring angles as such but used ideas about the ratios of the sides of triangles. These techniques would have been used to calculate lengths when designing and constructing buildings, such as the Giza Pyramids.

Chapter 7:
Pythagoras and trigonometry

In this chapter you will:

- use Pythagoras' theorem and the trigonometric ratios in complex 2D and 3D problems
- know and be able to prove the exact value of $\sin\theta$ and $\cos\theta$ for $\theta = 0°, 30°, 45°, 60°, 90°$ and for $\tan\theta$ for $\theta = 0°, 30°, 45°, 60°$
- use the sine and cosine rules in complex 2D and 3D problems
- know and use the area formula in complex 2D and 3D problems

Prior knowledge

- know and use Pythagoras' theorem and the trigonometric ratios
- know exact values of $\sin\theta$ and $\cos\theta$ for $\theta = 0°, 30°, 45°, 60°, 90°$ and for $\tan\theta$ for $\theta = 0°, 30°, 45°, 60°$
- know and use the sine and cosine rules
- know and use the formula for the area of a triangle

Chapter 7: Pythagoras and trigonometry

7.1 Pythagoras and trigonometry

For all right-angled triangles

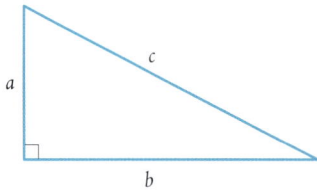

$a^2 + b^2 = c^2$ and

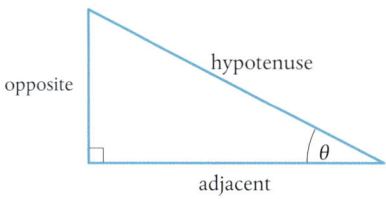

$$\sin\theta = \frac{\text{opposite}}{\text{hypotenuse}} \qquad \cos\theta = \frac{\text{adjacent}}{\text{hypotenuse}} \qquad \tan\theta = \frac{\text{opposite}}{\text{adjacent}}$$

Example 1

ABC and ACD are right-angled triangles.

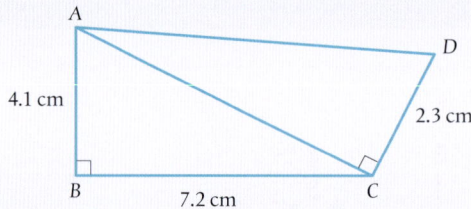

> 🗨 **Talking point**
> Why is it better to leave the length of AC in surd form?

Work out the size of angle ADC.

- Triangle ABC has two given lengths. Use Pythagoras. → $AC^2 = 4.1^2 + 7.2^2$

- Leave in surd form. → $AC = \sqrt{68.65}$ cm

- Triangle ACD now has two known lengths. Use the tan ratio. → $\tan ADC = \dfrac{\sqrt{68.65}}{2.3}$

  $ADC = \tan^{-1}\left(\dfrac{\sqrt{68.65}}{2.3}\right)$

- Give the answer correct to 3 s.f. → $ADC = 74.5°$

Chapter 7: Pythagoras and trigonometry

Example 2

ABCDEFGH is a cuboid.

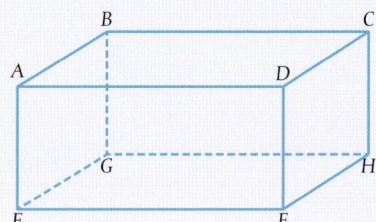

FE = 9.4 cm

FC = 17.3 cm

Angle FHE = 38°

Work out the length of CH.

$\sin 38 = \dfrac{9.4}{FH}$

$FH = \dfrac{9.4}{\sin 38} = 15.268\ldots$ cm — Visualise or sketch triangle FEH. Use the sine ratio.

$CH^2 = 17.3^2 - 15.268\ldots^2$ — Visualise or sketch triangle FCH. Use Pythagoras.

$CH = \sqrt{66.17\ldots}$

$CH = 8.13$ cm — Give the answer correct to 3 s.f.

Example 3

Prove that $\cos 45° = \dfrac{\sqrt{2}}{2}$.

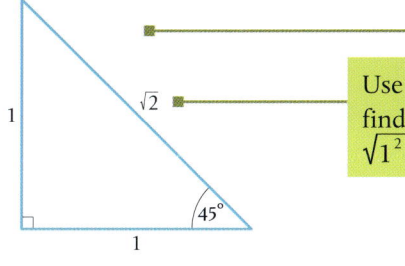

Start by drawing an isosceles right-angled triangle, with 2 sides of length 1.

Use Pythagoras to find the hypotenuse. $\sqrt{1^2 + 1^2} = \sqrt{2}$

$\cos \theta = \dfrac{\text{adjacent}}{\text{hypotenuse}}$ — Use the cosine ratio.

$\cos 45° = \dfrac{1}{\sqrt{2}}$ — Substitute the lengths from the diagram.

$\dfrac{1}{\sqrt{2}} \times \dfrac{\sqrt{2}}{\sqrt{2}} = \dfrac{\sqrt{2}}{2}$ — Rationalise the denominator.

So, $\cos 45° = \dfrac{\sqrt{2}}{2}$ — Complete the proof.

Chapter 7: Pythagoras and trigonometry

Practice

1 Find the lengths marked x.

a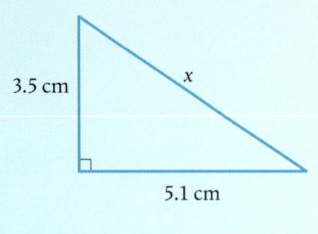
3.5 cm, 5.1 cm

b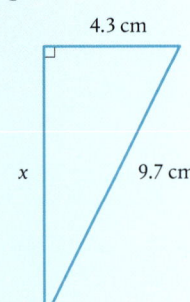
4.3 cm, 9.7 cm

c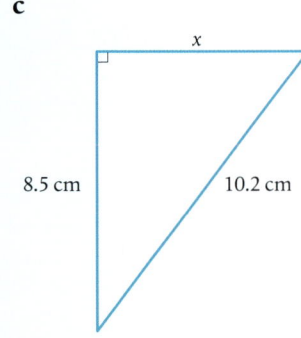
x, 8.5 cm, 10.2 cm

2 Find the lengths marked x

a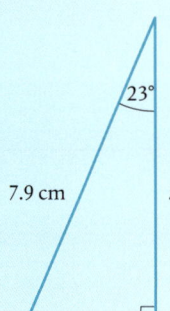
23°, 7.9 cm, x

b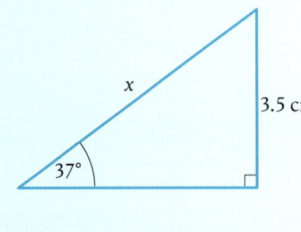
x, 3.5 cm, 37°

c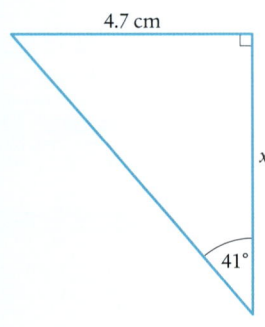
4.7 cm, x, 41°

3 Find the angles marked θ.

a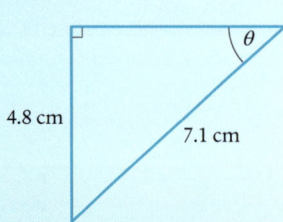
θ, 4.8 cm, 7.1 cm

b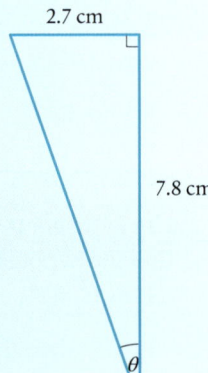
2.7 cm, 7.8 cm, θ

c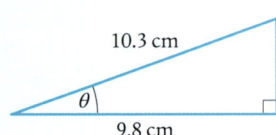
10.3 cm, θ, 9.8 cm

Chapter 7: Pythagoras and trigonometry

4 ▦ ABC and ACD are right-angled triangles.

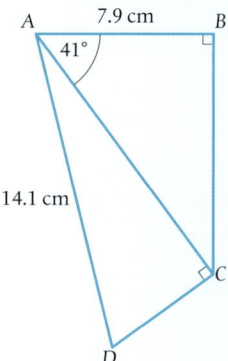

Work out the length of CD.

Hint for Q4
Work out the length of AC first.

5 **Reasoning** ABD and BCD are right-angled triangles.

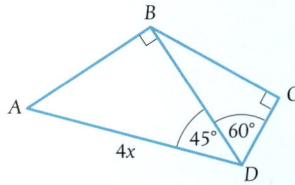

Work out an expression for the length of BC.

Give the answer in the form $x\sqrt{a}$, where a is an integer.

Hint for Q5
What is the exact value of cos 45°?

Exam-style question

6 ▦ ABC and ACD are right-angled triangles with side AC in common.

Angle ABC and angle ACD are right angles.

$AB = BC$
$AD = 3x$ cm
$CD = x$ cm

Find the perimeter of the quadrilateral ABCD. **(6 marks)**

7 ▦ ABCDEFGH is a cuboid.

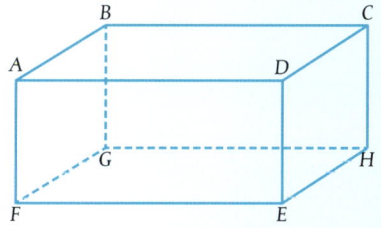

FH = 12.3 cm

Angle HFG = 54°   DE = 5.2 cm

Work out the size of angle EHD.

8 **Problem-solving** The diagram shows a cone, where the vertex A is directly above the centre of the base, O.

B is a point of the circumference of the base.

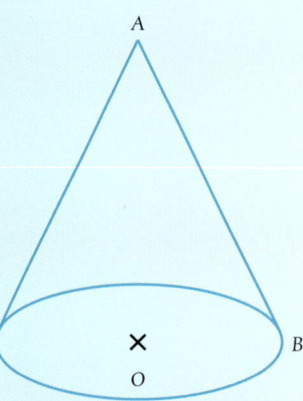

The volume of the cone is $\frac{147}{5}\pi$

The circumference of the base is 7π. Work out the size of angle ABO.

9 **Problem-solving** ABCDEFG is a triangular prism.

The cross section of the prism is an equilateral triangle of side length 4.

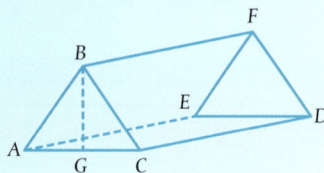

G is a point on the line AC such that angle AGB is a right angle.

The volume of the prism is $40\sqrt{3}$.

Work out the size of angle BGF.

10 **Problem-solving** A circle **C** has centre the origin.

Hint for Question 10

Start by drawing a diagram.

P is a point on the circumference of **C** with coordinates (3, 4).

Q is the point on the circumference of circle **C** with coordinates (q, 0) where q > 0.

R is the point with coordinates (3, 0).

Find the area of the shape PQR.

11 **Reasoning** Prove

a $\sin 30° = \frac{1}{2}$ b $\cos 30° = \frac{\sqrt{3}}{2}$ c $\tan 60° = \sqrt{3}$

Chapter 7: Pythagoras and trigonometry

7.2 Sine and cosine rules, and area of a triangle

For all triangles

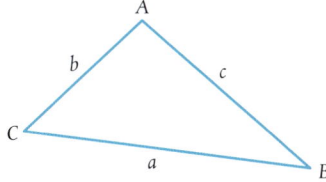

the sine rule	$\dfrac{a}{\sin A} = \dfrac{b}{\sin B} = \dfrac{c}{\sin C}$
or	$\dfrac{\sin A}{a} = \dfrac{\sin B}{b} = \dfrac{\sin C}{c}$
the cosine rule	$a^2 = b^2 + c^2 - 2bc \cos A$
the area of a triangle	area $= \dfrac{1}{2} ab \sin C$

Example 1

ABCD is a quadrilateral.

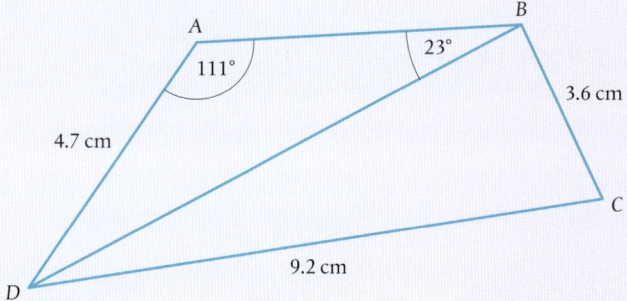

Work out the size of angle BCD.

$\dfrac{DB}{\sin DAB} = \dfrac{AD}{\sin ABD}$ — Triangle ABD has two given angles and one given side. Use the sine rule.

$\dfrac{DB}{\sin 111} = \dfrac{4.7}{\sin 23}$

$DB = \sin 111 \times \dfrac{4.7}{\sin 23}$ — Remember not to round prematurely. Keep the full decimal on your calculator.

$DB = 11.2... \text{ cm}$

$DB^2 = BC^2 + DC^2 - 2 \times BC \times DC \times \cos BCD$ — Triangle BCD now has three known lengths. Use the cosine rule.

$11.2...^2 = 3.6^2 + 9.2^2 - 2 \times 3.6 \times 9.2 \times \cos BCD$

$126.1... = 97.6 - 66.24 \cos BCD$

$\cos BCD = \dfrac{-28.5...}{66.24}$ — Partly evaluate and rearrange to make cos BCD the subject.

$BCD = \cos^{-1}\left(\dfrac{-28.5...}{66.24}\right)$

$BCD = 115°$ — Give the answer correct to 3 s.f.

131

Chapter 7: Pythagoras and trigonometry

Example 2

ABCD is a quadrilateral.

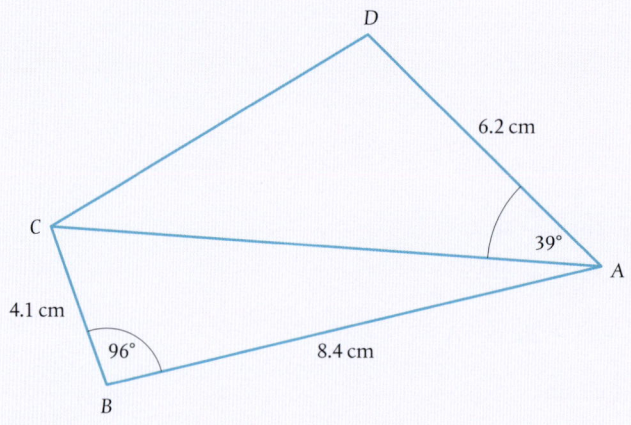

Work out the area of triangle ACD.

> Triangle ABC has two given lengths and one given angle. Use the cosine rule.

$AC^2 = AB^2 + BC^2 - 2 \times AB \times BC \times \cos ABC$

$AC^2 = 8.4^2 + 4.1^2 - 2 \times 8.4 \times 4.1 \times \cos 96$

> Remember not to round prematurely. Keep the full decimal on your calculator.

$AC = \sqrt{94.5\ldots}$

$= 9.72\ldots$ cm

> Use the area formula to find the area of triangle ACD.

area of $ACD = \frac{1}{2} \times AD \times AC \times \sin DAC$

$= \frac{1}{2} \times 6.2 \times 9.72\ldots \times \sin 39$

> Give the answer correct to 3 s.f.

$= 19.0$ cm²

Practice

1 Find the lengths marked x.

a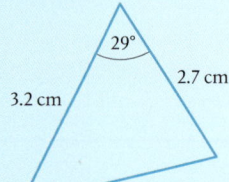

b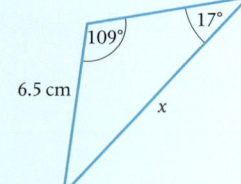

c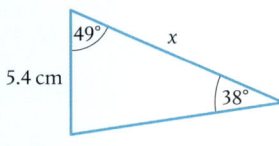

Chapter 7: Pythagoras and trigonometry

2 🖩 Find the angles marked θ.

a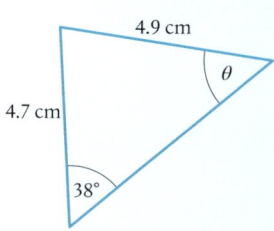

b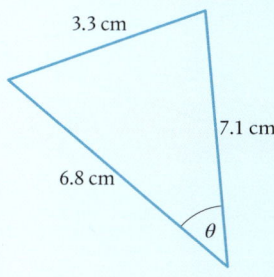

c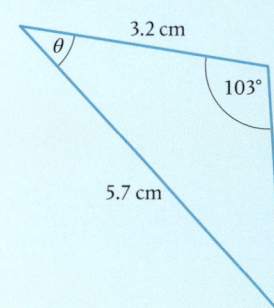

3 🖩 Work out the area of the triangles.

a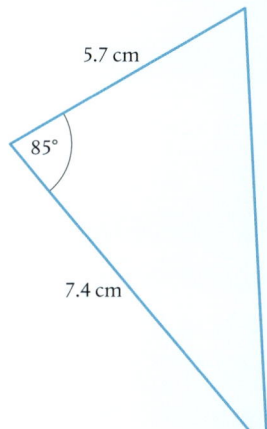

b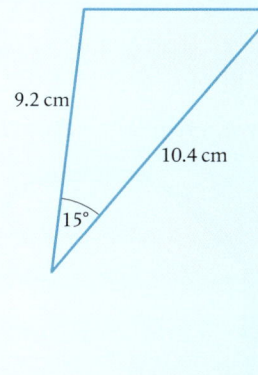

c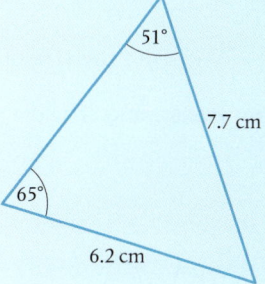

4 🖩 ABCD is a quadrilateral.

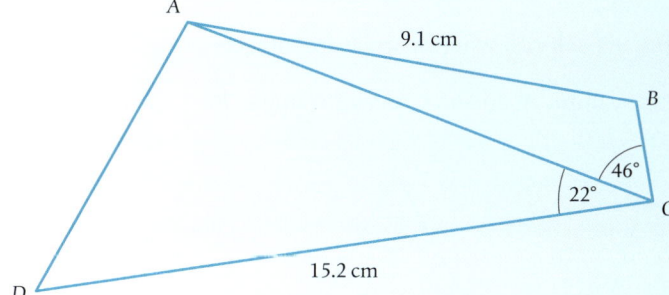

The area of triangle ACD is 34.2 cm².

Angle ABC > 90°

Work out the size of angle ABC.

5 ABCD is a quadrilateral.

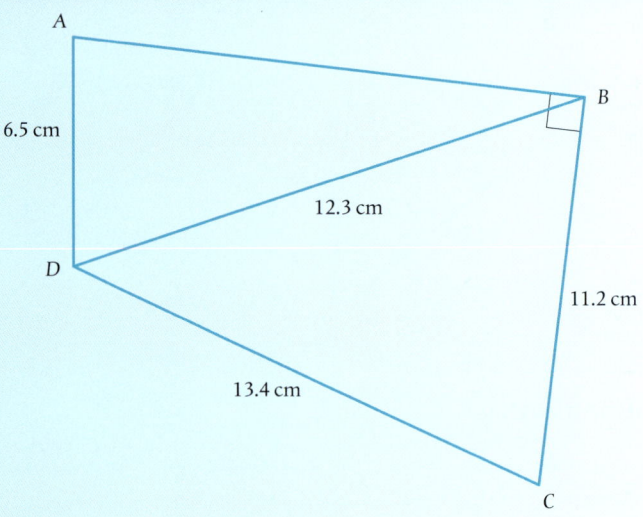

Work out the size of angle DAB.

Hint for Q6

Split the octagon into isosceles triangles.

6 **Reasoning** The diagram shows a regular octagon, inscribed in a circle with area 16π.

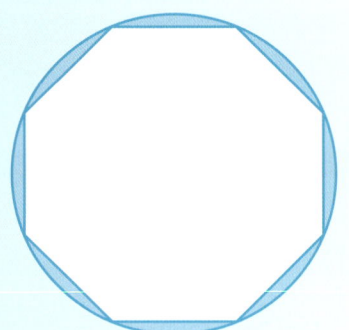

Work out the area of one of the shaded sections.

Give the answer in the form $a(\pi - b\sqrt{2})$, where a and b are integers.

Hint for Q7

Use the information in the question to form an equation.

7 The diagram shows a trapezium, A, and an isosceles triangle, B.

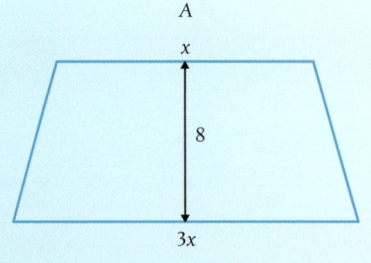 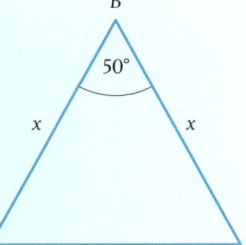

The ratio area of A : area of B = 2 : 5.

Find the value of x.

Chapter 7: Pythagoras and trigonometry

8 **Problem-solving** The diagram shows a circle with centre O and radius 6.7 cm.

P, Q and R are points on the circumference of the circle.

The arc length PQR = 4.5π cm

Work out the area of the segment PQR.

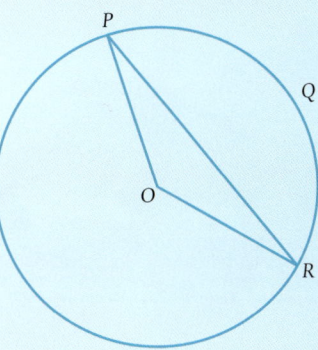

Exam-style question

9 The diagram shows a circle with centre O and radius 7.2 cm.

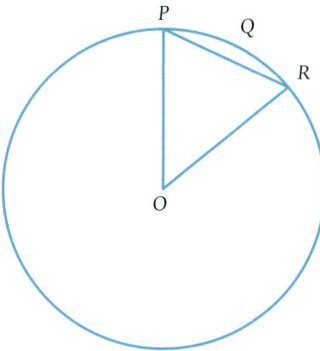

Points P, Q and R are on the circumference of the circle.

Length PR = 4.9 cm.

Work out the area of the segment PQR. **(6 marks)**

10 The diagram shows a triangular prism.

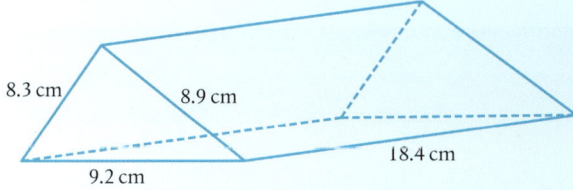

Work out the volume of the prism.

11 **Problem-solving** ABCD is the square base of a pyramid.

E is the vertex of the pyramid and is vertically above the centre of the base, O.

Angle AEB = 20°

Area of triangle AEB = 2.09 cm²

The ratio of length AB : length OE = 4 : 7

Work out the volume of the pyramid.

Hint for Q11

Draw and label a diagram with all the given information.

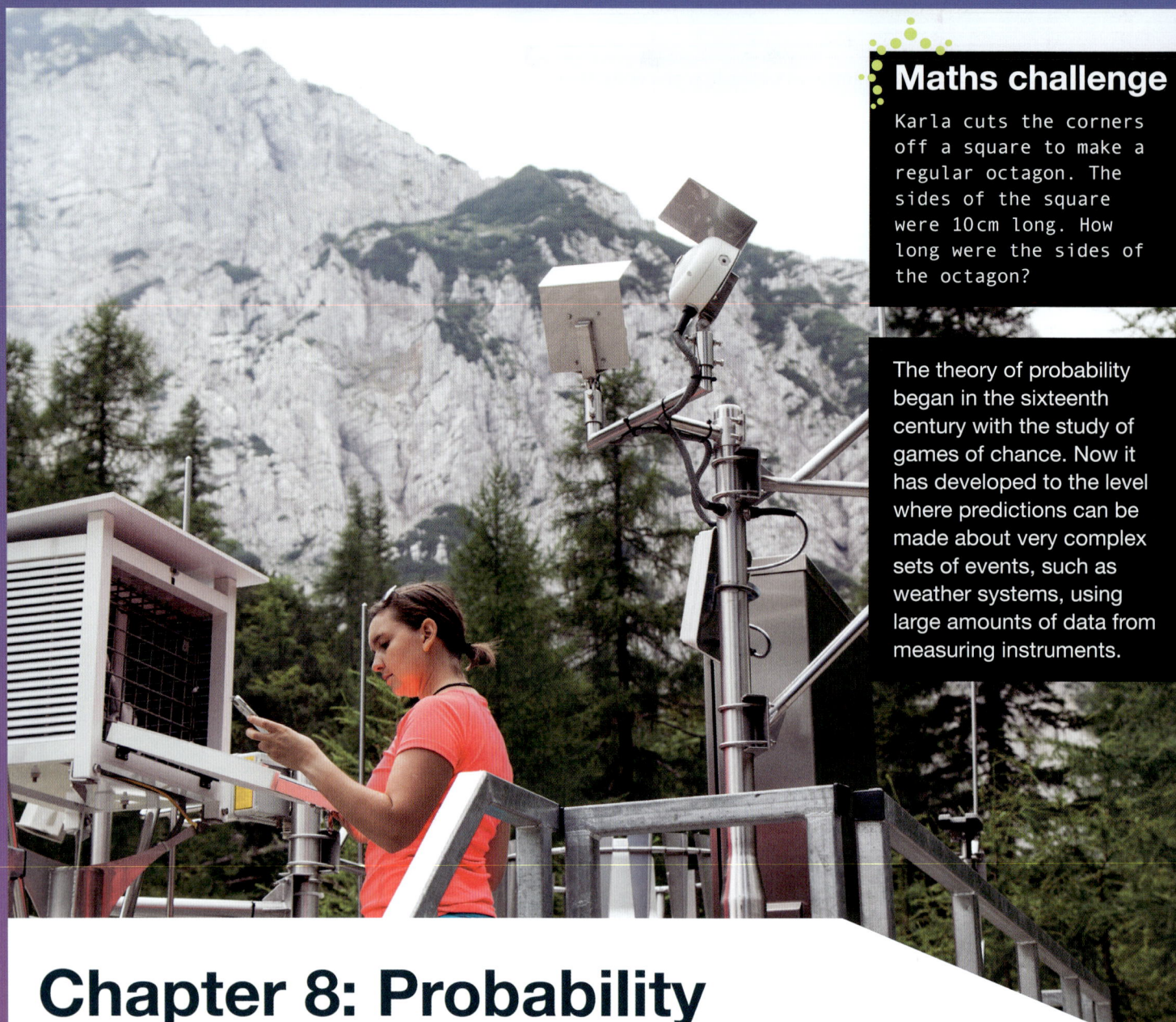

Maths challenge

Karla cuts the corners off a square to make a regular octagon. The sides of the square were 10 cm long. How long were the sides of the octagon?

The theory of probability began in the sixteenth century with the study of games of chance. Now it has developed to the level where predictions can be made about very complex sets of events, such as weather systems, using large amounts of data from measuring instruments.

Chapter 8: Probability

In this chapter you will:
- understand the notation and vocabulary of probability
- calculate the probability of independent and dependent combined events
- calculate conditional probabilities using various representations
- solve unstructured problems involving conditional probabilities

Prior knowledge
- solve quadratic equations
- understand and use set notation with Venn diagrams
- calculate probabilities of events from given information

Chapter 8: Probability

8.1 The language of probability

If all possible outcomes of an event are equally likely, then

$$\text{probability of an event} = \frac{\text{number of successful outcomes}}{\text{total number of possible outcomes}}$$

In some cases, the probabilities of different outcomes are not equal. If this is the case, it may be appropriate to carry out an experiment to estimate the probability of an event happening. In an experiment, the number of trials is recorded, as is the number of successful trials. As the number of trials increases, the **experimental** or **estimated probability** becomes nearer to the true value. In this case

$$\text{estimated probability of an event} = \frac{\text{number of successful trials}}{\text{total number of trials}}$$

Outcomes of an event that cannot happen at the same time are **mutually exclusive**.

- For two mutually exclusive events, A and B

  $P(A \text{ or } B) = P(A) + P(B)$

  This is the addition law for mutually exclusive events.

- For a set of mutually exclusive events, the sum of the probabilities = 1

  So, $P(A) + P(\text{not } A) = 1$
  Or, $P(\text{not } A) = 1 - P(A)$

When the outcome of one event does not affect the outcome of another, the two events are **independent**.

- For two independent events A and B

  $P(A \text{ and } B) = P(A) \times P(B)$

  This is the multiplicative law for independent events.

- For two sets, A and B, use the following notation

  Union (in A or in B or in both) = $A \cup B$

  Intersection (in A and in B) = $A \cap B$

  Complement (not in A) = A'

- Use Venn diagrams to represent sets visually

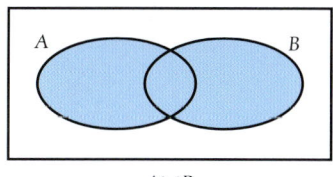
$A \cup B$

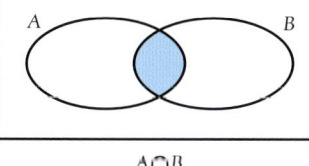
$A \cap B$

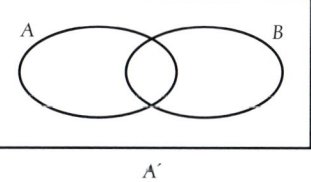
A'

> **Talking point**
> Why are estimated probabilities different to theoretical probabilities?
> Will they ever be the same?

Chapter 8: Probability

Example 1

150 Year 10 and Year 12 students must study Geography or History.

39 of the 93 Year 10 students study Geography.

A total of 82 students study History.

a Use this information to complete a two-way table.

b One of the students is chosen at random.

  i What is the probability this student is in Year 12 and studies History?

  ii Given the student studies History, what is the probability this student was in Year 10?

 Talking point

What other diagram would be suitable to display and analyse this data?

Insert the information given in the question into the table. Use this information to work out the missing values.

a

	Geography	History	Total
Year 10	39	93 − 39 = 54	93
Year 12	68 − 39 = 29	82 − 54 = 28	150 − 93 = 57
Total	150 − 82 = 68	82	150

b **i** $\frac{28}{150}$

  ii $\frac{54}{82}$

- Number of Year 10 who chose History.
- Total number of students who chose History.
- Number of Year 12 students who study History.
- Total number of students.

Chapter 8: Probability

Example 2

$\xi = \{1, 2, 3, 4, 5, 6, 7, 8, 9\}$

$A = \{\text{odd numbers}\}$

$B = \{\text{multiples of 3}\}$

a Use this information to complete a Venn diagram.

b One number is chosen at random from ξ.

    i Work out the probability that this number is in $A \cap B$.

    ii Work out the probability that this number is in B'.

a

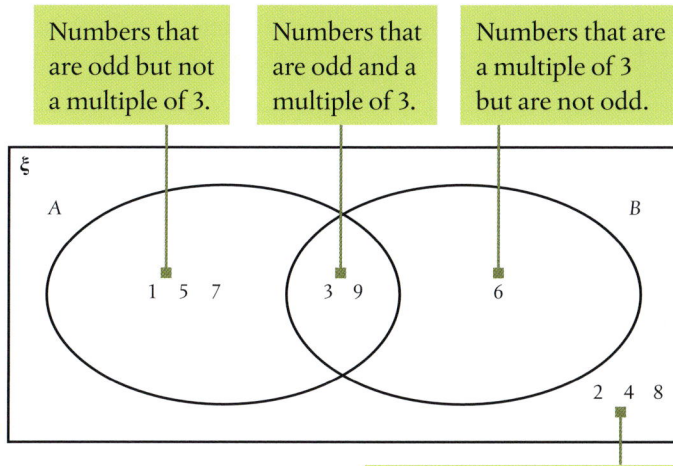

Numbers that are odd but not a multiple of 3.

Numbers that are odd and a multiple of 3.

Numbers that are a multiple of 3 but are not odd.

Numbers that are not odd and are not multiples of 3.

b i $A \cap B = (\text{in } A \text{ and in } B) = 3 \text{ and } 9$

   so

   $P(A \cap B) = \dfrac{2}{9}$

   Number of elements of $A \cap B$.

   Number of elements of ξ.

ii $B' = (\text{not in } B) = 1, 2, 4, 5, 7, 8$

   so

   $P(B') = \dfrac{6}{9}$

   Number of elements of B'.

   Number of elements of ξ.

Practice

1 Some students went on an activity day.

The students were either in Year 7 or Year 8.

They chose one activity from art or music or drama.

Of the 41 students, 15 chose music.

8 of the 30 students from Year 7 chose art.

None of the Year 8 students chose art.

An equal number of Year 7 and Year 8 students chose drama.

a Use this information to complete the two-way table.

	Art	Music	Drama	Total
Year 7				
Year 8				
Total				

One of the students was chosen at random.

b Work out the probability that this student was in Year 7 and chose music.

c Given the student chose drama, work out the probability that they were in Year 8.

2 Jonas has a biased dice.

When someone throws the dice, the probability that it lands on a 6 is $\frac{1}{3}$.

Jonas is going to throw the dice twice.

Hint for Q2a

P(not A) = 1 − P(A)

a Complete the probability tree diagram.

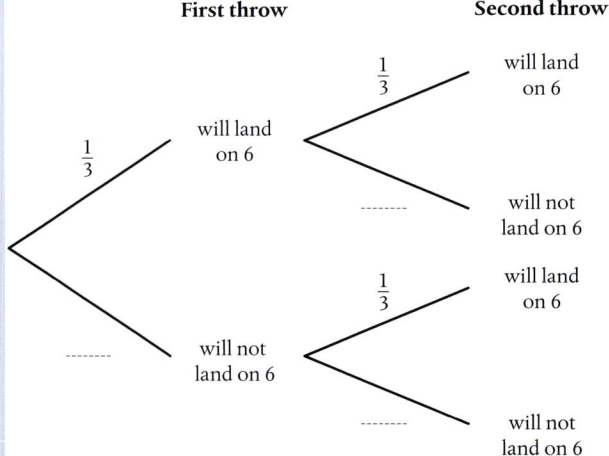

b Work out the probability that Jonas throws 6 both times.

c Work out the probability that Jonas throws 6 exactly once.

Hint for Q2b
P(A and B) = P(A) × P(B)

Hint for Q2c
P(A or B) = P(A) + P(B)

3 A bag contains 3 red balls, 5 blue balls and 2 green balls.

Brenda is going to choose 1 ball at random from the bag, record the colour and replace it back in the bag.

She will then repeat this a second time.

a Use this information to complete the probability tree diagram.

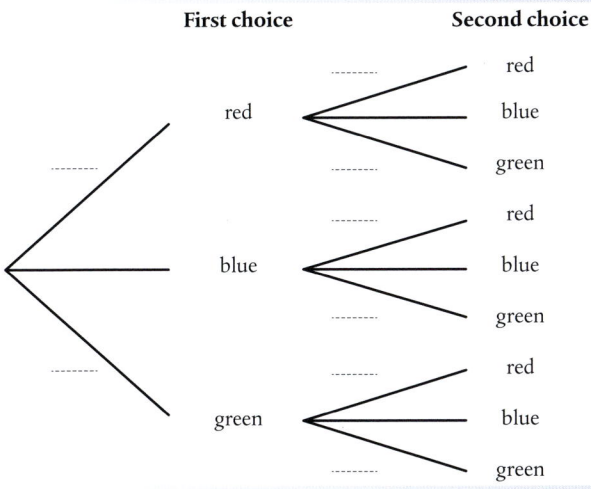

b Work out the probability that Brenda chooses two balls of the same colour.

4 Reasoning There are 4 red counters, 5 white counters and 1 blue counter in a bag.

Berat is trying to work out the probability that the counter is red or blue, when one counter is chosen at random from the bag. Here is his working:

P(red or blue) = P(red) × P(blue)

$= \frac{4}{10} \times \frac{1}{10} = \frac{4}{100}$

What is wrong with Berat's working?

5 ξ = {1, 2, 3, 4, 5, 6, 7, 8, 9}

A = {even numbers}

B = {factors of 6}

a Use this information to complete the Venn diagram.

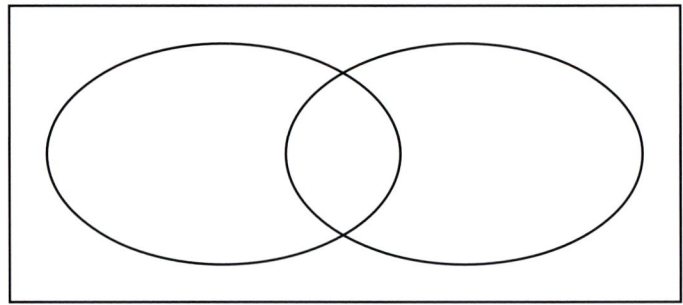

One number is chosen at random. Work out

b P(A∩B)

c P(A∪B)

d P(A')

6 A survey asks 120 people if they prefer athletics, boxing or cycling.

30 of the people were children, the rest were adults.

12 of the children and 26 of the adults preferred boxing.

35 of the 45 people who chose athletics were adults.

a **Problem-solving** Work out the number of adults who preferred cycling.

One of the people is chosen at random.

b i Work out the probability that they were a child who preferred cycling.

ii Given they preferred boxing, work out the probability that they were an adult.

7 A market research company asked 40 men and 60 women if they go to the gym.

35% of the men said yes.

42% of the women said yes.

a Draw a tree diagram to show this information.

One of the people from the survey is chosen at random.

b Work out the probability that this person does not go to the gym.

Chapter 8: Probability

8 **Problem-solving** A researcher asked 100 people if they shop at Supermarket A or at Supermarket B.

67 said supermarket A.

23 said supermarket B.

13 said supermarket A and B.

Work out $P(A \cup B)$.

Hint for Q8

Draw a Venn diagram.

9 **Problem-solving** Arun has a collection of 250 stamps.

A = {English stamps}

B = {Stamps over 50 years old}

157 of the stamps are English.

89 of the stamps are more than 50 years old.

$P(A \cap B) = 0.064$

Work out $P(A \cup B)'$.

Exam-style question

10 A shopkeeper puts the vinyl in his music shop into sections.

The sections are rock, jazz, popular and classical.

The shopkeeper puts each vinyl into one of these sections.

a Draw a Venn diagram to represent this situation.
A vinyl is selected at random.
The probability that this vinyl is from the popular section is 0.3 **(2 marks)**

b Write down the value of
P(rock) + P(jazz) + P(classical). **(1 mark)**

11 Here is a Venn diagram for two events, A and B, where $P(A) \neq 0$ and $P(B) \neq 0$.

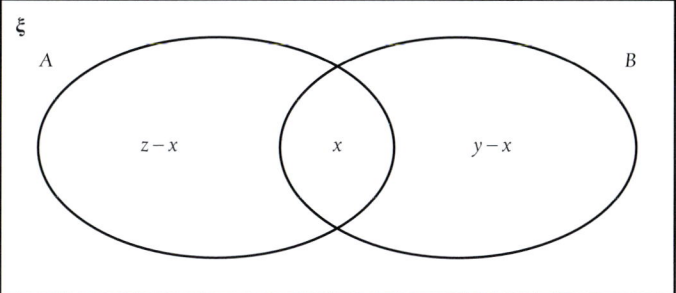

The number of elements in the universal set is n.

Use this information to show that
$P(A \cup B) = P(A) + P(B) - P(A \cap B)$. **(4 marks)**

Chapter 8: Probability

8.2 Probability problems

Conditional probability is the probability of one event happening depending on a previous event having already happened.

Write the probability of event A happening **given that** event B has already happened as:

P(A|B)

If two events are not independent, the outcome of one affects the other, so:

P(A∩B) = P(B|A) × P(A)

Example 1

A bag contains 5 blue balls and 7 red balls.

A ball is taken at random from the bag, and **not** replaced.

A second ball is then taken at random.

a Draw a probability tree diagram for this information.

b Work out the probability that the two balls are the same colour.

a

This is the probability of choosing a blue on the second choice **given that** blue was chosen first.

Probability of blue is 5 out of a total of 12.

Probability of red is 7 out of a total of 12.

This is the probability of choosing a red on the second choice **given that** blue was chosen first.

First choice → Second choice

- $\frac{5}{12}$ blue
  - $\frac{4}{11}$ blue
  - $\frac{7}{11}$ red
- $\frac{7}{12}$ red
  - $\frac{5}{11}$ blue
  - $\frac{6}{11}$ red

This is the probability of choosing a red on the second choice **given that** blue was chosen first.

This is the probability of choosing a blue on the second choice **given that** red was chosen first.

P(A∩B) = P(B|A) × P(A)

P(A or B) = P(A) + P(B)

b P(both blue) = P(blue|blue) × P(blue) = $\frac{4}{11} \times \frac{5}{12} = \frac{20}{132}$

P(both red) = P(red|red) × P(red) = $\frac{6}{11} \times \frac{7}{12} = \frac{42}{132}$

P(both same colour) = P((both blue) ∩ (both red))
= P((both blue) + (both red))
= $\frac{20}{132} + \frac{42}{132} = \frac{62}{132}$

Chapter 8: Probability

Example 2

There are only green counters and blue counters in a bag.

There are 5 more blue counters than green counters.

There are more than 50 counters in the bag.

Klara is going to take 2 counters from the bag.

The probability that both counters are the same colour is $\frac{49}{99}$

Work out how many green counters are in the bag.

Let x be the number of green counters,

then $x + 5$ is the number of blue counters,

and the total number of counters is $2x + 5$,

then P(both green) $= \frac{x-1}{2x+4} \times \frac{x}{2x+5} = \frac{x^2 - x}{4x^2 + 18x + 20}$ — $P(A \cap B) = P(B|A) \times P(A)$

P(both blue) $= \frac{x+4}{2x+4} \times \frac{x+5}{2x+5} = \frac{x^2 + 9x + 20}{4x^2 + 18x + 20}$

P(both same colour) = P(both green) + P(both blue) — $P(A \text{ or } B) = P(A) + P(B)$

$= \frac{x^2 - x + x^2 + 9x + 20}{4x^2 + 18x + 20}$

$= \frac{2x^2 + 8x + 20}{4x^2 + 18x + 20}$

$\frac{2x^2 + 8x + 20}{4x^2 + 18x + 20} = \frac{49}{99}$

$99(2x^2 + 8x + 20) = 49(4x^2 + 18x + 20)$ — Form a quadratic equation and rearrange it ready to solve.

$198x^2 + 792x + 1980 = 196x^2 + 882x + 980$

$2x^2 - 90x + 1000 = 0$

$x^2 - 45x + 500 = 0$

Factorising the LHS gives

$(x - 25)(x - 20) = 0$ — Factorise.

$x = 25$ or $x = 20$

If $x = 20$, there are 20 green counters and, therefore, there are 25 blue and a total of 45. But the question tells us there are *more than* 50, so the number of green counters must be 25.

 Talking point

Why does this example not include a tree diagram? Would a tree diagram help answer the question?

Chapter 8: Probability

Practice

1 The probability that it will snow on a given day in December is 0.1.

If it does snow, the probability that Brian's train will be late is 0.62.

If it does not snow, the probability that Brian's train will not be late is 0.92.

    a Draw a probability tree diagram for this information.

    b Work out the probability that, on a given day in December, it **does not** snow and Brian's train **is late**.

2 There are 5 blue balls and 4 red balls in a bag.

Two balls are taken at random from the bag.

    a Draw a tree diagram to show this information

    b Work out the probability that both balls will be the same colour.

3 There are 7 red counters and 2 blue counters in a bag.

One counter is taken at random from the bag, and not replaced.

Then 4 green counters are added to the bag.

A second counter is then taken from the bag.

    a Draw a tree diagram to show this information.

    b Work out the probability that both counters are different colours.

4 **Problem-solving** There are 12 sweets in a jar.

x of the sweets are lemon flavour. The rest of the sweets are strawberry flavour.

Grace takes, at random, two sweets from the jar.

Find an expression, in terms of x, for the probability that Grace takes one sweet of each flavour.

> **Hint for Q4**
> Draw a tree diagram.

5 There are x grey T-shirts and 5 black T-shirts in a bag.

Suha takes, at random, two T-shirts from the bag.

The probability that Suha takes one grey T-shirt and one black T-shirt is $\frac{20}{39}$

    a **Reasoning** Show that $2x^2 - 21x + 40 = 0$.

    b Work out the probability that Suha takes two grey T-shirts.

> **Hint for Q5b**
> Solve the equation to find the value of x.

6 **Problem-solving** In a bag there are 7 blue counters, 3 green counters and y red counters.

Carlos takes, at random, two counters from the bag.

Given that the probability that both counters are green is 0.01, work out the probability that Carlos takes one blue counter and one red counter.

Chapter 8: Probability

7 **Problem-solving** Daisy and Enlai are both often late for school.

A teacher recorded whether they are late on a number of days.

ξ = {total number of days recorded}

D = {number of days Daisy is late}

E = {number of days Enlai is late}

The teacher drew this Venn diagram using their results.

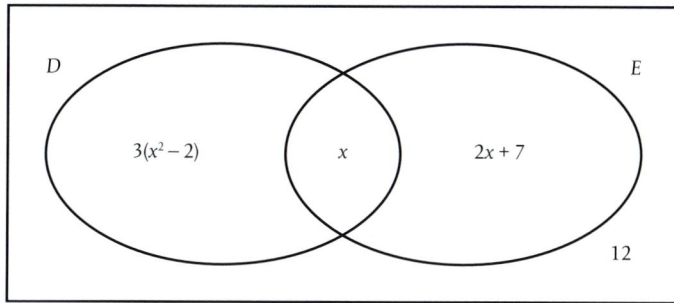

The teacher selects one day, at random, from the survey.

Given that on that day Daisy was late, the probability that Enlai was also late is $\frac{1}{18}$.

Work out how many days the teacher recorded.

Exam-style question

8 A bag contains x buttons.

There are 6 red buttons in the bag.

The rest of the buttons are white.

Jo takes two buttons at random from the bag.

The probability that both buttons are the same colour is $\frac{17}{35}$.

Find the value of x. **(6 marks)**

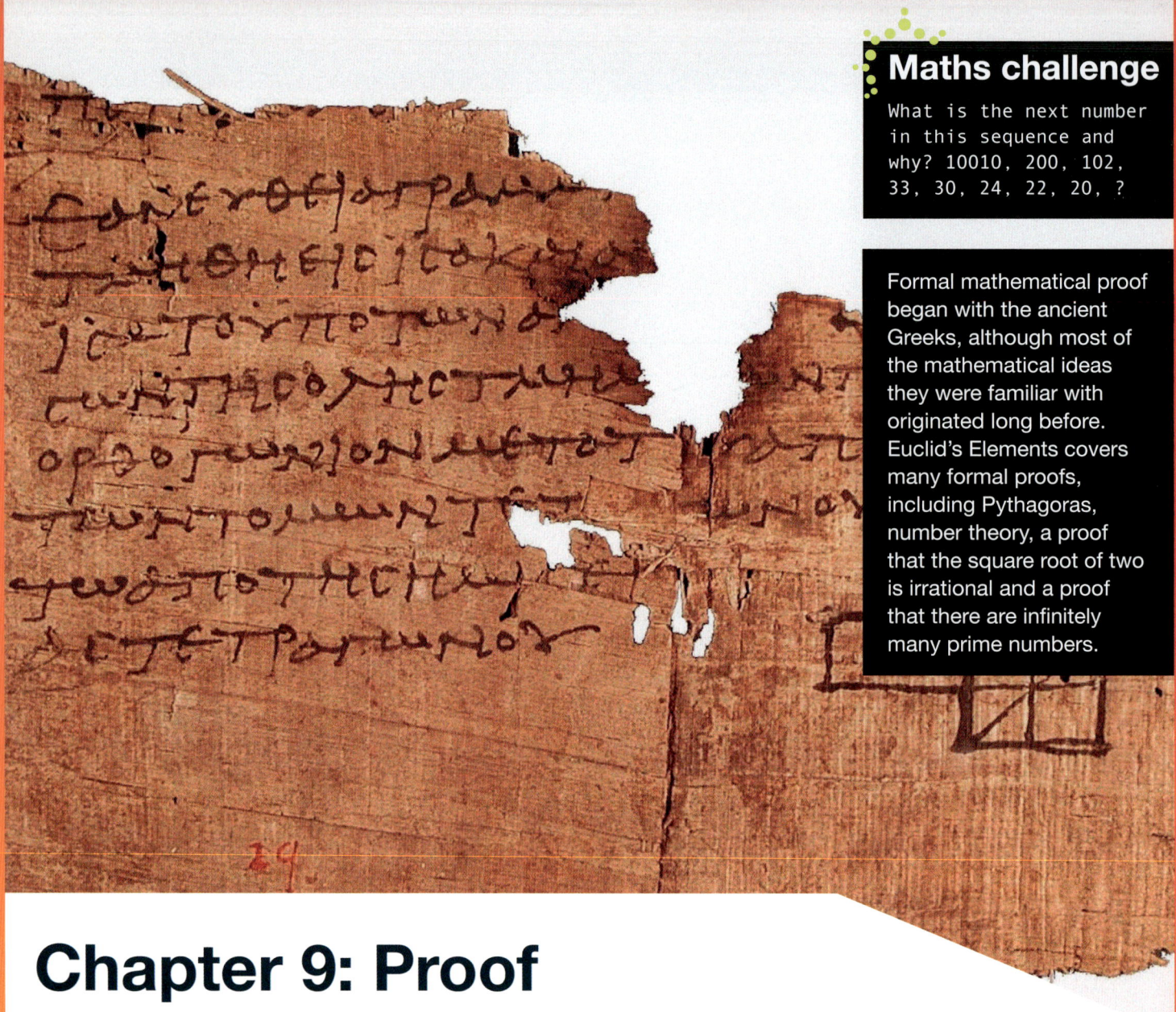

Maths challenge

What is the next number in this sequence and why? 10010, 200, 102, 33, 30, 24, 22, 20, ?

Formal mathematical proof began with the ancient Greeks, although most of the mathematical ideas they were familiar with originated long before. Euclid's Elements covers many formal proofs, including Pythagoras, number theory, a proof that the square root of two is irrational and a proof that there are infinitely many prime numbers.

Chapter 9: Proof

In this chapter you will:
- understand and use the structure of mathematical proof, proceeding from given assumptions through a series of logical steps to a conclusion
- use different methods of proof
  - proof by deduction
  - proof by exhaustion
  - disproof by counter example
  - geometric proof

Prior knowledge
- expand polynomials
- factorise quadratic expressions
- complete the square
- find the turning point of a quadratic curve algebraically
- rationalise a denominator
- simplify algebraic fractions
- use the trigonometric ratios
- identify perpendicular lines algebraically
- understand the coordinate geometry of a circle
- know and use the angle properties of a circle

Chapter 9: Proof

9.1 Proof by deduction

A mathematical proof is a logical and structured argument to show that a statement is true in all cases.

To prove a statement by **deduction**, start by stating known facts or theorems, then work through a series of logical steps to a final statement that confirms that the statement is true.

When completing proofs, it is often necessary to define odd and even integers.

Use $2n$ to represent an even number for some integer n (because all even numbers are multiples of 2).

Use $2n + 1$ or $2n - 1$ to represent an odd number (one more or one less than an even number).

Use $n, n + 1, n + 2, \ldots$ to represent consecutive numbers.

Example 1

Prove that $(x + a)^3 \equiv x^3 + 3x^2a + 3xa^2 + a^3$

$(x + a)^3 = (x + a)(x + a)(x + a)$

$\qquad = (x^2 + ax + ax + a^2)(x + a)$

$\qquad = (x^2 + 2ax + a^2)(x + a)$

$\qquad = x^3 + ax^2 + 2ax^2 + 2a^2x + a^2x + a^3$

$\qquad = x^3 + 3x^2a + 3xa^2 + a^3$

Therefore, $(x + a)^3 \equiv x^3 + 3x^2a + 3xa^2 + a^3$

Start with the left-hand side and work logically to reach the right-hand side.

The final line of a proof is a statement to show what has been proved.

Alternatively, use the fourth row of Pascal's triangle, 1, 3, 3, 1.

$(x + a)^3 = x^3a^0 + 3x^2a^1 + 3x^1a^2 + x^0a^3$

$\qquad = x^3 + 3x^2a + 3xa^2 + a^3$

Therefore, $(x + a)^3 \equiv x^3 + 3x^2a + 3xa^2 + a^3$

 Talking point
Which method do you think is best? Would it be the same if the power of the expansion changed?

Example 2

Prove that the function $y = 2x^2 - 12x + 25$ is positive for all values of x.

$2x^2 - 12x + 25 = 2(x^2 - 6x) + 25$

$\qquad = 2\big((x - 3)^2 - 9\big) + 25$

$\qquad = 2(x - 3)^2 + 7$

Complete the square.

Therefore, the turning point of the function is $(-3, 7)$. This is a positive quadratic, so the turning point is a minimum.

Therefore, the range of $y = 2x^2 - 12x + 25$ is $y \geq 7$ and so is always positive.

Chapter 9: Proof

Example 3

Here is a right-angled triangle.

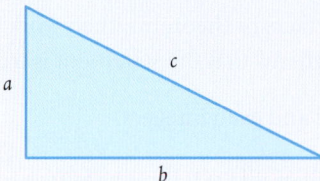

Four of these triangles enclose the white square below.

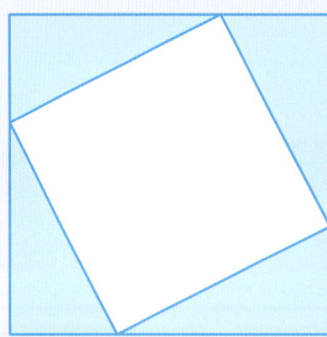

Use this diagram to prove Pythagoras' theorem.

Pythagoras' theorem states that $a^2 + b^2 = c^2$.

The area of the white square is c^2.

The proof must show that the area of the white square can also be written as $a^2 + b^2$.

(area of white square) = (area of large square) − (area of 4 triangles)

$$(\text{area of large square}) = (a + b)^2$$
$$= a^2 + 2ab + b^2$$
$$(\text{area of 4 triangles}) = 4 \times \left(\frac{1}{2} a \times b\right)$$
$$= 2ab$$

So

(area of white square) = (area of large square) − (area of 4 triangles)
$$= (a^2 + 2ab + b^2) - (2ab)$$
$$= a^2 + b^2$$

Therefore, for any right-angled triangle $a^2 + b^2 = c^2$.

 **Talking point**

How do you know that the area of the white square is c^2?

Chapter 9: Proof

Practice

1. Prove algebraically that the difference between the squares of any two consecutive integers is equal to the sum of these two integers.

2. **Reasoning** The product of two consecutive positive integers is added to the larger of these two integers.

   Prove that the result is always a square number.

3. Prove algebraically that $(2n + 1)^2 - (2n + 1)$ is even for all positive values of n.

4. Prove that the sum of the squares of three consecutive odd numbers is always 11 more than a multiple of 12.

   Hint for Q4
   The standard expression for an odd number is $2n + 1$.

5. Here are the first 5 terms of an arithmetic sequence

   3, 7, 11, 15, 19

   Prove that the difference between the squares of two consecutive terms of this sequence is always a multiple of 8.

   Hint for Q5
   Find the nth term of the sequence first.

6. Prove that $\dfrac{x}{1 + \sqrt{12}} \equiv \dfrac{2x\sqrt{3} - x}{11}$.

7. Prove that the product of any two odd numbers is odd.

8. **Problem-solving** Prove that $3x^2 + 24x + 56$ is positive for all values of x.

9. **Problem-solving** Prove that $2x^2 - 6x + \dfrac{11}{2}$ is positive for all values of x.

10. **Problem-solving** Prove that $10x - 5x^2 - 12$ is negative for all values of x.

11. Prove that $\left(x + \dfrac{1}{x}\right)^3 \equiv x^3 + 3x + \dfrac{3}{x} + \dfrac{1}{x^3}$.

    Hint for Q11
    Here are the first five rows of Pascal's triangle:

    ```
            1
          1   1
        1   2   1
      1   3   3   1
    1   4   6   4   1
    ```

12. Prove that, for all positive values of n, $\dfrac{(n + 2)^2 - (n + 1)^2}{2n^2 + 3n} = \dfrac{1}{n}$.

13. The diagram shows a triangle ABC, with base BC and perpendicular height h.

    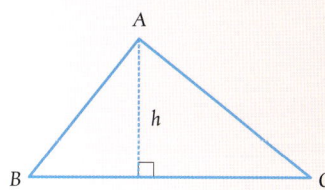

    Use this diagram to prove $\dfrac{b}{\sin B} = \dfrac{c}{\sin C}$

Chapter 9: Proof

14 The diagram shows a triangle ABC, with base c and perpendicular height $CD = h$.

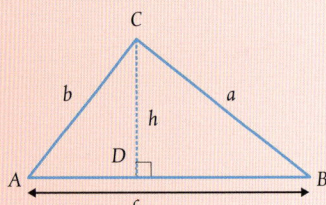

- **a** Given that $AD = x$, use Pythagoras to write an expression for b^2 in terms of h and x.
- **b** Use Pythagoras to write an expression for a^2 in terms of c, x and h.
- **c** Hence, write an expression for b^2 in terms of a, c and x.
- **d** Using trigonometry, write an expression for x in terms of b and A.
- **e** Hence, prove that $a^2 = b^2 + c^2 - 2bc \cos A$.

Exam-style question

Hint for Q15

Draw a diagram and label the sides and angles a, b, c and A, B, C.

15 Using area of a triangle $= \frac{1}{2}$ base $\times$ height, prove that

area of a triangle $= \frac{1}{2} ab \sin C$ **(3 marks)**

Chapter 9: Proof

9.2 Proof by exhaustion, and disproof by counter example

When a statement has a limited number of possible cases, it is possible to prove the statement by **exhaustion**. This means proving it is true for every case.

Example 1

Prove that all square numbers are either a multiple of 4 or one more than a multiple of 4.

There are two possible cases: when the number is even and when the number is odd. — *Deal with each case separately.*

Even numbers:

$(2n)^2 = 4n^2$ — *Use $2n$ to represent an even number.*

$4n^2$ is a multiple of 4

Odd numbers:

$(2n + 1)^2 = 4n^2 + 4n + 1$ — *Use $2n + 1$ to represent an odd number.*

$ = 4(n^2 + n) + 1$

$4(n^2 + n)$ is a multiple of 4, so

$4(n^2 + n) + 1$ is one more than a multiple of 4.

Since all integers are even or odd, this proves that all square numbers are either a multiple of 4 or one more than a multiple of 4. — *Don't forget to write a statement to complete the proof.*

Disproof by counter example means disproving a mathematical statement by giving a single example that does not work.

Example 2

Given an example to disprove the statement
'The product of any two prime numbers is always odd.'

2 and 3 are both prime numbers.

$2 \times 3 = 6$

6 is not odd, so the statement is not true. — *Don't forget to write a statement to complete the proof.*

Practice

1. **Reasoning** Show that $n^2 + 2$ is not a multiple of 4 when $3 \leq n \leq 7$.

   Hint for Q1
   Use proof by exhaustion using each possible value of n in turn.

2. Given that n is a prime number such that $3 < n < 14$, show that $(n - 1)(n + 1)$ is a multiple of 12.

3. Given that n is an odd integer such that $2 < n < 20$, show that n is either prime or a product of exactly 2 primes.

4. Show that $n^2 - 1$ is a multiple of 3 if n is not a multiple of 3.

Chapter 9: Proof

Hint for Q5
If n is 2 more than a multiple of 5 it must be in the form $5n + 2$.

5 **Problem-solving** Given that n and m are consecutive integers with $n < m$, and n is 2 more than a multiple of 5, prove than nm is 1 more than a multiple of 5.

6 Given that $3 \leq n \leq 7$, prove that n^3 is either a multiple of 9 or one more or one less than a multiple of 9.

7 **Problem-solving** Prove that, if n is 1 more or 1 less than a multiple of 4, then n^2 is always 1 more than a multiple of 4.

8 **Problem-solving** Given that n is odd, prove that n^4 is always 1 more than a multiple of 8.

Hint for Q8
Here are the first five rows of Pascal's triangle:
```
        1
       1 1
      1 2 1
     1 3 3 1
    1 4 6 4 1
```

9 Prove that no square numbers end in an 8.

10 **Reasoning** Give an example to disprove each of these statements.

a The square of an integer is always even.

b The sum of two numbers is always greater than both numbers.

c All of the digits in the product of 2 even numbers are always even.

d $x^2 - x + 7$ is prime for all integer values of x.

e Given $a > b$, then $a^2 > b^2$.

f Given that n is prime, $n + 2$ is always prime.

Exam-style question

11 Given that n is an integer such that $1 \leq n \leq 5$, prove that $n^2 + n + 11$ is prime. **(3 marks)**

12 Disprove the statement: $n^2 - n + 3$ is prime for all integer values of n. **(2 marks)**

Chapter 9: Proof

9.3 Geometric proof

Geometric proof uses logical geometric reasoning to prove a statement or theorem about geometry.

A diagram is vital to any geometric proof. Use it to display any given information, then work through a series of logical steps to reach the given statement.

Always state each step of the proof, even if it repeats information from the question, along with evidence to support that statement. It is important not to miss out any steps.

Example 1

Prove that $A(1, 2)$, $B(5, 5)$ and $C(8, 1)$ are the vertices of a right-angled triangle.

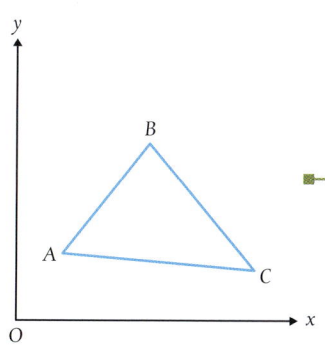

Sketch a diagram to show the information in the question.

gradient of $AB = \dfrac{5-2}{5-1} = \dfrac{3}{4}$

gradient of $BC = \dfrac{1-5}{8-5} = -\dfrac{4}{3}$

gradient of $CA = \dfrac{2-1}{1-8} = -\dfrac{1}{7}$

gradient of AB × gradient of $BC = \dfrac{3}{4} \times \left(-\dfrac{4}{3}\right) = -1$

So, AB and BC are perpendicular and the triangle is right-angled.

Talking point
Could you also prove this using Pythagoras?

Chapter 9: Proof

Example 2

ABCD is a rhombus.

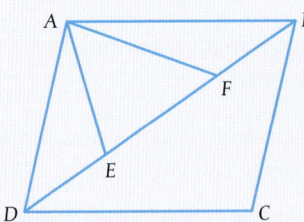

E and F are points on the line BD such that BF = ED.

Prove that triangle AED is congruent to triangle AFB.

Give evidence to support each statement.

BF = ED   (given information)
AD = AB   (sides of a rhombus are equal in length)
∠ADE = ∠ABF   (since AD = AB, triangle ABD is isosceles and base angles of an isosceles triangle are equal)

Therefore, triangle AED is congruent to triangle AFB using Side-Angle-Side (SAS).

Example 3

Prove that the angle in a semi-circle is always 90°.

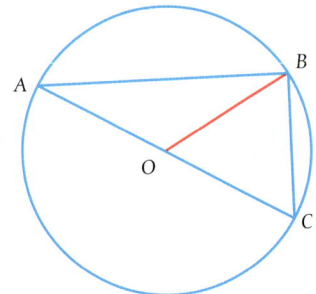

Draw a suitable diagram and label any relevant points. Refer to them in the proof.

Triangles AOB and BOC are isosceles since OA, OB and OC are radii of the circle.

Let ∠AOB = x

Then,

∠OAB = ∠OAB = $\frac{180 - x}{2}$ = $90 - \frac{x}{2}$ (base angles of an isosceles triangle are equal)

∠BOC = 180 − x   (angles on a straight line sum to 180°)

∠OBC = ∠OCB = $\frac{180 - (180 - x)}{2}$ = $\frac{x}{2}$ (base angles of an isosceles triangle are equal)

So, ∠ABC = ∠OBA + ∠OBC

= $\left(90 - \frac{x}{2}\right) + \frac{x}{2}$

= 90

Therefore, the angle in a semi-circle always equals 90°.

Chapter 9: Proof

Practice

1 Prove that $A(1, -1)$, $B(4, 2)$ and $C(7, -1)$ are the vertices of a right-angled isosceles triangle.

2 Prove that $A(-1, 2)$, $B(1, 3)$, $C(3, 2)$ and $D(1, -2)$ are the vertices of a kite.

3 Prove that $A(-2, 1)$, $B(-1, 3)$, $C(4, 3)$ and $D(3, 1)$ are the vertices of a parallelogram.

4 **Reasoning** ABCD is a parallelogram.

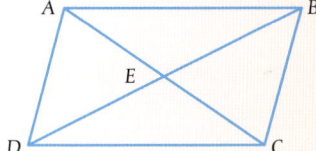

E is the point where the two diagonals intersect.

Prove the triangle ABE is congruent to triangle DCE.

5 **Reasoning** ABCD is a quadrilateral.

$AB = DC$

Angle $ABC =$ Angle BCD

Prove that $AC = BD$

6 **Problem-solving** Prove that $3y - 4x + 15 = 0$ is a tangent to the circle with equation $x^2 + 2x + y^2 - 4x - 20 = 0$.

Hint for Q6
First find the point of intersection between the circle and the tangent.

7 The diagram shows a circle, centre O.

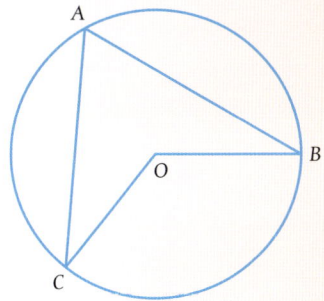

Hint for Q7
Draw in the line AO to form two isosceles triangles.

A, B and C are points on the circle.

Prove that angle CAB is half the size of angle COB.

8 The diagram shows a circle, centre O.

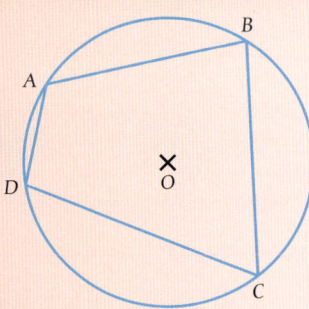

A, B, C and D are points on the circle.

Prove that opposite angles in a cyclic quadrilateral sum to 180°.

9 The diagram shows a circle, centre O.

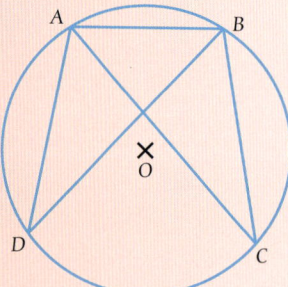

A, B, C and D are points on the circle.

Prove that angle DAC = Angle DBC.

Exam-style question

10 Prove that the angle between a chord and a tangent is equal to the angle subtended by the chord in the alternate segment.

The answer must include evidence, including any relevant circle theorems.

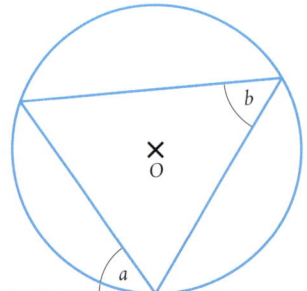

(4 marks)

Maths challenge

What is the largest number that cannot be made by adding together 5s and 7s?

Vectors can be used to describe positions and movements in two- and three-dimensional space but they can also be used in more dimensions to describe related sets of data. One example of this is in machine learning programs for streaming media, describing sets of user preferences as vectors and performing operations on them. This can be used to track viewing and make recommendations.

Chapter 10: Vectors

In this chapter you will:
- add and subtract position vectors and multiply a vector by a scalar
- calculate the distance between two points represented by position vectors
- use vectors to solve problems in pure mathematics and in context

Prior knowledge
- calculate exactly with surds and simplify surd expressions involving squares
- simplify and manipulate algebraic expressions
- use Pythagoras' theorem
- use the trigonometric ratios in right-angled triangles
- use the sine rule and cosine rule
- use the area formula for a triangle

Chapter 10: Vectors

10.1 Position vectors

To add or subtract two position vectors, add or subtract the *x*-terms and add or subtract the *y*-terms.

For the two position vectors $\mathbf{a} = \begin{pmatrix} x_1 \\ y_1 \end{pmatrix}$ and $\mathbf{b} = \begin{pmatrix} x_2 \\ y_2 \end{pmatrix}$

$\mathbf{a} + \mathbf{b} = \begin{pmatrix} x_1 + x_2 \\ y_1 + y_2 \end{pmatrix}$ and $\mathbf{a} - \mathbf{b} = \begin{pmatrix} x_1 - x_2 \\ y_1 - y_2 \end{pmatrix}$.

To multiply a position vector by a scalar, multiply each value by the scalar

$4\mathbf{a} = \begin{pmatrix} 4 \times x_1 \\ 4 \times y_1 \end{pmatrix}$ and $-5\mathbf{b} = \begin{pmatrix} -5 \times x_2 \\ -5 \times y_2 \end{pmatrix}$

For the two position vectors on the grid, *x* is the distance between them.

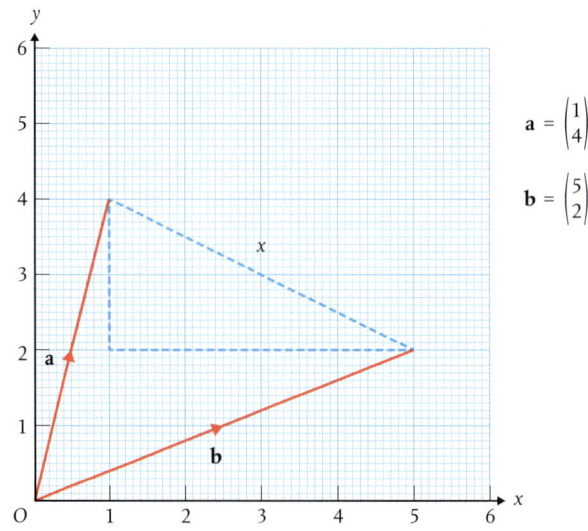

To find the distance between the vectors, create a right-angled triangle with hypotenuse *x* and use Pythagoras' theorem

$x = \sqrt{(2-4)^2 + (5-1)^2}$

$x = \sqrt{20}$

$x = 4.47$ (2 d.p.)

In general

for $\mathbf{a} = \begin{pmatrix} x_1 \\ y_1 \end{pmatrix}$ and $\mathbf{b} = \begin{pmatrix} x_2 \\ y_2 \end{pmatrix}$

distance $= \sqrt{(x_2 - x_1)^2 + (y_2 - y_1)^2}$

> **Talking point**
> Does the order in which you subtract matter? Justify your answer.

Chapter 10: Vectors

Example 1

$a = \begin{pmatrix} 4 \\ -1 \end{pmatrix}$ and $b = \begin{pmatrix} -3 \\ 6 \end{pmatrix}$

Work out $2a - 3b$.

$$2a - 3b = \begin{pmatrix} 8 \\ -2 \end{pmatrix} - \begin{pmatrix} -9 \\ 18 \end{pmatrix}$$

$$= \begin{pmatrix} 8 - -9 \\ -2 - 18 \end{pmatrix}$$

$$= \begin{pmatrix} 17 \\ -20 \end{pmatrix}$$

$2a = \begin{pmatrix} 2 \times 4 \\ 2 \times -1 \end{pmatrix} = \begin{pmatrix} 8 \\ -2 \end{pmatrix}$ and
$3b = \begin{pmatrix} 3 \times -3 \\ 3 \times 6 \end{pmatrix} = \begin{pmatrix} -9 \\ 18 \end{pmatrix}$

Example 2

$a = \begin{pmatrix} 3 \\ 2 \end{pmatrix}$ and $b = \begin{pmatrix} 2 \\ -4 \end{pmatrix}$

Given that $3a - c = 2b$, find c.

$3a - c = 2b$

$c = 3a - 2b$

$= \begin{pmatrix} 9 \\ 6 \end{pmatrix} - \begin{pmatrix} 4 \\ -8 \end{pmatrix}$

$= \begin{pmatrix} 9 - 4 \\ 6 - -8 \end{pmatrix}$

$= \begin{pmatrix} 5 \\ 14 \end{pmatrix}$

Rearrange the equation to make c the subject.

$3a = \begin{pmatrix} 3 \times 3 \\ 3 \times 2 \end{pmatrix} = \begin{pmatrix} 9 \\ 6 \end{pmatrix}$ and
$2b = \begin{pmatrix} 2 \times 2 \\ 2 \times -4 \end{pmatrix} = \begin{pmatrix} 4 \\ -8 \end{pmatrix}$

Substitute values into the rearranged equation, and calculate.

Example 3

The position vector a has column vector $\begin{pmatrix} 7 \\ 3 \end{pmatrix}$.

The position vector b has column vector $\begin{pmatrix} -3 \\ -1 \end{pmatrix}$.

Work out the distance between a and b.

$\text{distance} = \sqrt{(-3 - 7)^2 + (-1 - 3)^2}$

$= \sqrt{116}$

$= 10.77$ (2 d.p.)

Use Pythagoras' theorem.

Chapter 10: Vectors

Practice

1. $a = \begin{pmatrix} 2 \\ 1 \end{pmatrix}$ $b = \begin{pmatrix} 3 \\ 7 \end{pmatrix}$ $c = \begin{pmatrix} 6 \\ 9 \end{pmatrix}$

   Work out

   a $b + c$
   b $a - c$
   c $a + (b - c)$

2. $d = \begin{pmatrix} 4 \\ 6 \end{pmatrix}$ $e = \begin{pmatrix} -2 \\ 3 \end{pmatrix}$ $f = \begin{pmatrix} 1 \\ -5 \end{pmatrix}$

   Work out

   a $2d + 4e$
   b $2f - d$
   c $4e + (2d - 3f)$

3. $g = \begin{pmatrix} -9 \\ 2 \end{pmatrix}$ $h = \begin{pmatrix} 5 \\ -4 \end{pmatrix}$

   a Given that $g + j = h$, find j.

   b Given that $2h + k = 2g$, find k.

   c Given that $3g - 2m = 2h$, find m.

4. The position vector **a** has column vector $\begin{pmatrix} 4 \\ 2 \end{pmatrix}$. The position vector **b** has column vector $\begin{pmatrix} 7 \\ 9 \end{pmatrix}$. Find the distance between **a** and **b**. Give the answer correct to 2 decimal places.

5. **Reasoning** The position vector **c** has column vector $\begin{pmatrix} 8 \\ -1 \end{pmatrix}$. The position vector **d** has column vector $\begin{pmatrix} 6 \\ -7 \end{pmatrix}$. Find the distance between **c** and **d**. Give the answer in the form $a\sqrt{b}$ where a and b are integers.

6. **Reasoning** The position vector **c** has column vector $\begin{pmatrix} -3 \\ -7 \end{pmatrix}$. The position vector **d** has column vector $\begin{pmatrix} -6 \\ -13 \end{pmatrix}$. Find the distance between **c** and **d**. Give the answer in the form $a\sqrt{b}$ where a and b are integers.

Hint for Q7
Draw a diagram to help visualise the problem.

7. **Problem-solving** Three position vectors, **a**, **b** and **c**, form the three vertices of a triangle.

   $a = \begin{pmatrix} 3 \\ 8 \end{pmatrix}$ $b = \begin{pmatrix} 3 \\ 2 \end{pmatrix}$ $c = \begin{pmatrix} 9 \\ 3 \end{pmatrix}$

   a Work out the perimeter of the triangle. Give the answer correct to 2 decimal places.

   b Work out the size of the angle between **ab** and **bc**. Give the answer correct to 2 decimal places.

Chapter 10: Vectors

8 **Problem-solving** The diagram shows a triangle OAB.

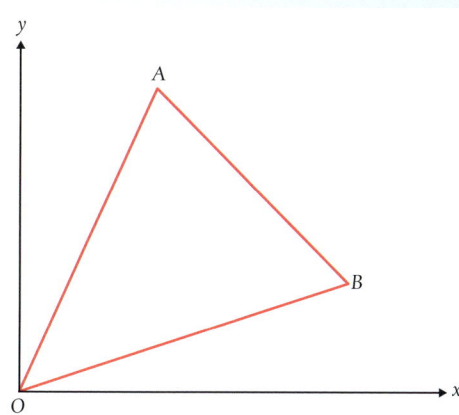

The column vector $\begin{pmatrix} 4 \\ 11 \end{pmatrix}$ represents A.

The column vector $\begin{pmatrix} 10 \\ 3 \end{pmatrix}$ represents B.

Work out the area of OAB. Give the answer correct to 2 decimal places.

Hint for Q8
Use right-angled triangles to find the size of the angle between OA and the x-axis, and between OB and the x-axis.

9 **Problem-solving** The diagram shows part of a circle **C** with equation $x^2 + y^2 = 25$.

Hint for Q9
What is the length of OA?

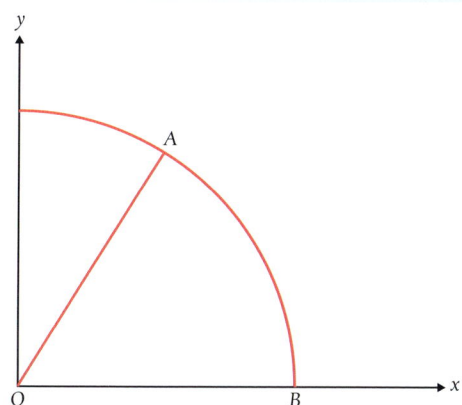

The column vector $\begin{pmatrix} x \\ 4 \end{pmatrix}$ represents A.

Work out the area of sector OAB to 2 d.p.

Exam-style question

10 The point E has coordinates (−3, 4), with respect to the origin.

The point F is such that $\overrightarrow{EF}$ is $\begin{pmatrix} -6 \\ 7 \end{pmatrix}$.

Express, as a column vector, the position vector **F**. **(2 marks)**

11 The position vector **W** has column vector $\begin{pmatrix} -2 \\ 3 \end{pmatrix}$.

The position vector **Z** has column vector $\begin{pmatrix} x \\ -5 \end{pmatrix}$.

The distance between the two position vectors is 10 units.

Work out the possible values of x. **(4 marks)**

Chapter 10: Vectors

10.2 Solving geometric problems

The result of adding and subtracting vectors is a **resultant vector**.

The order of addition does not matter: **a** + **b** = **b** + **a**.

Parallel vectors are multiples of each other. For example, **a**, 4**a**, 7**a**, $\frac{1}{2}$**a**, −4**a**, −$\frac{7}{9}$**a** are all parallel because they are all multiples of **a**.

To prove two vectors are parallel, show that they are multiples of the same vector or resultant vector.

To prove that two vectors are colinear (that they form a single straight line), prove that the two vectors

- are parallel
- have at least one point in common.

Example 1

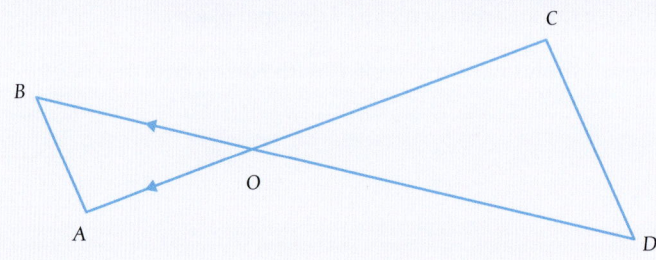

$\overrightarrow{OA}$ = **a** $\qquad\qquad\qquad\qquad$ $\overrightarrow{OB}$ = **b**

OC = 3 × OA $\qquad\qquad\qquad$ OD = 3 × OB

a Work out, in terms of **a** and **b**

  i $\overrightarrow{OC}$ $\qquad\qquad$ **ii** $\overrightarrow{OD}$ $\qquad\qquad$ **iii** $\overrightarrow{AB}$

b Prove that $\overrightarrow{AB}$ and $\overrightarrow{DC}$ are parallel.

a i $\overrightarrow{OC} = -3 \times \overrightarrow{OA}$

$\qquad\;\; = -3\mathbf{a}$

— $\overrightarrow{OC}$ is three times the length of $\overrightarrow{OA}$ and in the opposite direction.

ii $\overrightarrow{OD} = -3 \times \overrightarrow{OB}$

$\qquad\;\; = -3\mathbf{b}$

— $\overrightarrow{OD}$ is three times the length of $\overrightarrow{OB}$ and in the opposite direction.

iii $\overrightarrow{AB} = \overrightarrow{AO} + \overrightarrow{OB}$

$\qquad\;\; = -\mathbf{a} + \mathbf{b}$

$\qquad\;\; = \mathbf{b} - \mathbf{a}$

— $\overrightarrow{AO} + \overrightarrow{OB}$ and $\overrightarrow{OB} + \overrightarrow{AO}$ give the same resultant.

b $\overrightarrow{DC} = \overrightarrow{DO} + \overrightarrow{OC}$

$\qquad\;\; = 3\mathbf{b} + -3\mathbf{a}$

$\qquad\;\; = 3\mathbf{b} - 3\mathbf{a} = 3(\mathbf{b} - \mathbf{a})$

— Find $\overrightarrow{DC}$.

Therefore, $\overrightarrow{DC} = 3\overrightarrow{AB}$ and they are parallel.

Chapter 10: Vectors

Example 2

OABC is a trapezium such that AB and OC are parallel.

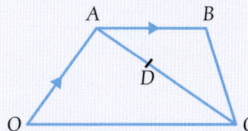

$\overrightarrow{OA} = \mathbf{a}$

$\overrightarrow{AB} = \mathbf{b}$

$\overrightarrow{OC} = 2\overrightarrow{AB}$

D is a point on $\overrightarrow{AC}$ such that

AD : DC = 1 : 2

Prove that ODB is a straight line.

$\overrightarrow{AC} = \overrightarrow{AO} + \overrightarrow{OC}$ — First find $\overrightarrow{AC}$.

$\phantom{\overrightarrow{AC}} = -\mathbf{a} + 2\mathbf{b} = 2\mathbf{b} - \mathbf{a}$

$\overrightarrow{OD} = \overrightarrow{OA} + \overrightarrow{AD}$ — To prove ODB is a straight line, find $\overrightarrow{OD}$ and $\overrightarrow{DB}$.

$\phantom{\overrightarrow{OD}} = \overrightarrow{OA} + \frac{1}{3}\overrightarrow{AC}$

$\phantom{\overrightarrow{OD}} = \mathbf{a} + \frac{1}{3}(2\mathbf{b} - \mathbf{a})$

$\phantom{\overrightarrow{OD}} = \frac{2}{3}(\mathbf{a} + \mathbf{b})$

$\overrightarrow{DB} = \overrightarrow{DA} + \overrightarrow{AB}$

$\phantom{\overrightarrow{DB}} = -\frac{1}{3}\overrightarrow{AC} + \overrightarrow{AB}$

$\phantom{\overrightarrow{DB}} = -\frac{1}{3}(2\mathbf{b} - \mathbf{a}) + \mathbf{b}$

$\phantom{\overrightarrow{DB}} = \frac{1}{3}(\mathbf{a} + \mathbf{b})$

So, $\overrightarrow{OD} = 2\overrightarrow{DB}$ which shows they are parallel.

$\overrightarrow{OD}$ and $\overrightarrow{DB}$ also have point D in common, so they are colinear; that is, ODB is a straight line.

> 💬 **Talking point**
> The answer to this example used $\overrightarrow{OD}$ and $\overrightarrow{DB}$. Is it possible to use a different vector in place of one of these to complete the proof?

To complete the proof, state what the algebra demonstrates.

Practice

1. In this diagram $\overrightarrow{OA} = 3\mathbf{a} + \mathbf{b}$, $\overrightarrow{OB} = 4\mathbf{b} - \mathbf{a}$ and $\overrightarrow{OC} = -3\mathbf{a} - 5\mathbf{b}$.

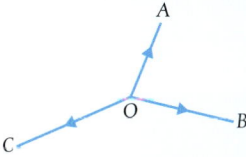

Write each vector in terms of **a** and **b** in its simplest form.

a $\overrightarrow{BC}$ b $\overrightarrow{AB}$ c $\overrightarrow{CA}$

Chapter 10: Vectors

2 O is the centre of the regular hexagon ABCDEF.

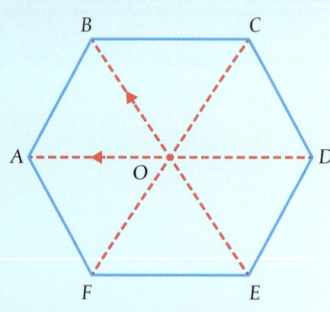

$\vec{OA} = \mathbf{a}$ $\vec{OB} = \mathbf{b}$

Write each vector in terms of **a** and **b** in its simplest form.

a $\vec{BC}$ b $\vec{AB}$ c $\vec{CA}$

d $\vec{FC}$ e $\vec{EB}$ f $\vec{FB}$

3 **Reasoning** In the diagram, N and M are the midpoints of OA and OB respectively.

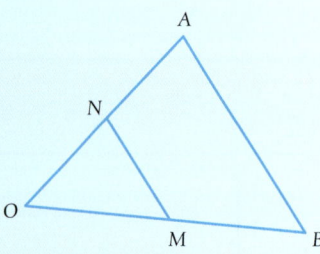

$\vec{OA} = \mathbf{a}$ $\vec{OB} = \mathbf{b}$

a Write $\vec{AB}$ in terms of **a** and **b**.

b Prove that $\vec{AB}$ and $\vec{NM}$ are parallel.

4 **Reasoning** OBCE is a parallelogram.

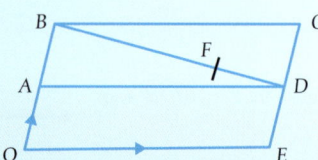

A and D are the midpoints of OB and EC respectively.

$\vec{OA} = \mathbf{a}$ $\vec{OE} = \mathbf{e}$

F is a point on BD such that $BF:FD = 2:1$

a Find, in terms of **a** and **e**, $\vec{BD}$.

b Prove that OFC is a straight line.

Hint for Q4

Find an expression for $\vec{OF}$ and for $\vec{FC}$

Chapter 10: Vectors

5 **Problem-solving** In the diagram ABC is a straight line.

M is the midpoint of OA, and B is the midpoint of AC.

N is a point on the line OB.

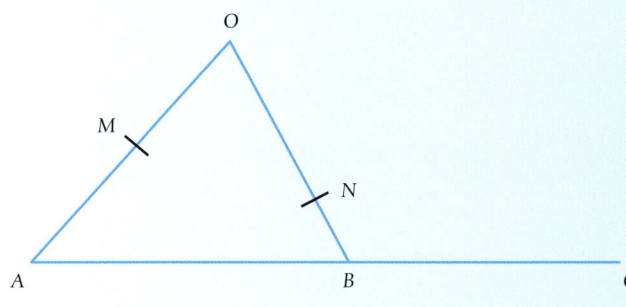

$\overrightarrow{OA} = \mathbf{a}$ $\overrightarrow{OB} = \mathbf{b}$

Given that MNC is a straight line, find ON : NB.

6 **Problem-solving** In the diagram, OPM and APN are straight lines.

M is the midpoint of AB.

P is the point on OM such that OP : PM = 4 : 1.

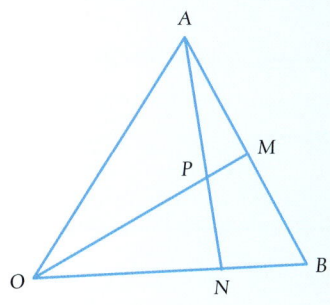

$\overrightarrow{OA} = \mathbf{a}$ $\overrightarrow{OB} = \mathbf{b}$

Work out ON : NB.

7 **Reasoning** OAB is a triangle where P, M and N are the midpoints of OA, AB and BO, respectively.

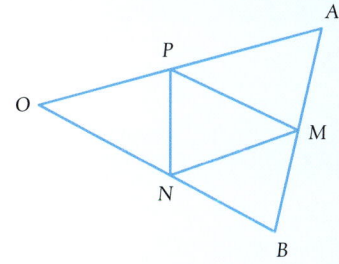

$\overrightarrow{OP} = \mathbf{a}$ $\overrightarrow{PN} = \mathbf{b}$

Prove that triangle APM is congruent to triangle PMN.

Hint for Q7

Remember the conditions for congruence, SAS and ASA.

Chapter 10: Vectors

8 Problem-solving *OABC* is a rectangle, where *M* is the midpoint of *OA* and *BQ* : *QC* = 1 : 4.

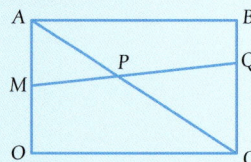

AC and MQ intersect at the point *P*.

$\vec{OA} = 2\mathbf{a}$ $\vec{OC} = 4\mathbf{c}$

Prove $\vec{PC} = \frac{8}{13}\vec{AC}$.

9 Problem-solving *ABC* is a triangle.

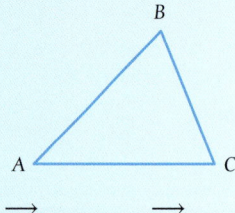

$\vec{AB} = \mathbf{a}$ $\vec{AC} = \mathbf{c}$

M is the midpoint of *AB*

D is a point such that *BCD* forms a straight line and $\vec{BC} : \vec{CD} = 3 : 2$.

E is a point such that $\vec{DE} = \lambda\mathbf{a}$, and *MCE* is a straight line.

Work out the value of λ.

Exam-style question

10 *ABC* is a triangle.

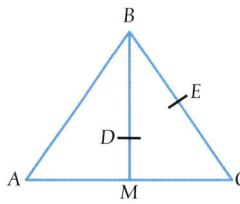

$\vec{BA} = \mathbf{x}$

$\vec{BC} = \mathbf{y}$

M is the midpoint of *AC*.

E is the midpoint of *BC*.

D is the point on *BM* such that *BD* : *DM* is 2 : 1.

Write *ED* in terms of **x** and **y**.

Give the answer in its simplest form. **(7 marks)**

Mixed Practice

1 Simplify

   a $(8a^9c^{15})^{\frac{2}{3}}$ **b** $\left(\dfrac{x^4}{9x^6y^8}\right)^{-\frac{3}{2}}$

Exam-style question

2 Without using a calculator, simplify $\dfrac{\sqrt{45} + \sqrt{720}}{\sqrt{3}}$.

Give the answer in the form $\sqrt{a}$ where a is an integer. **(4 marks)**

3 Expand and simplify $x(2x-3)(5x-4)(2x+3)$.

4 a Find an equation of the straight line that passes through the points with coordinates $(9, -2)$ and $(-6, 8)$.

  b Find an equation of the line that is perpendicular to the line in part **a** and passes through the point with coordinates $(6, 6)$.

5 Simplify

  a $\dfrac{m^9 - m^7}{m^4}$ **b** $\dfrac{(2a^3b^5)^4 \times 9a^2b}{3a^4b^5}$

6 Reasoning Find the value of n such that $4^3 \times 2^n = 16$.

Exam-style question

7 Given that $\dfrac{175x^3 - 112x}{15x^2 + 2x - 8} \times \left(\dfrac{1}{5x-4} + \dfrac{1}{x}\right) = n$.

Find the value of n. **(5 marks)**

8 **Problem-solving** The diagram shows triangle PQT.

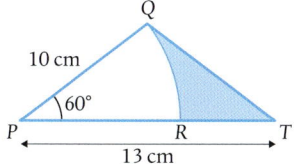

PQR is a sector of a circle with centre P.

Find the area of the shaded part of the diagram bounded by the straight lines QT, RT and the curve QR.

9 Problem-solving The points A, B and C are such that B is on a bearing of $070°$ from A and C is on a bearing of $140°$ from B.

The distance AB is 4.3 metres and the distance BC is 5.6 metres.

  a Find, by calculations, the distance AC.

  b Find, by calculations, the bearing of A from C.

Mixed Practice

Exam-style question

10 Without using a calculator and showing all the stages of working, write $\dfrac{\sqrt{18}+\sqrt{98}}{\sqrt{5}}$ in the form $\sqrt{n}$ where n is an integer. **(3 marks)**

11 **Reasoning** Draw a sketch of the graph with equation $y = (x-4)(x+6)(2x+7)$.

Show the coordinates of any points at which the graph intercepts with the coordinate axes.

12 **Reasoning** Disprove, by counter example, the following statement.

For any values of m and n, if $m > n$, then $m^2 + m > n^2 + n$

Exam-style question

13 The graph of $y = pq^x$ passes through the points with coordinates $(0, 7)$ and $(3, 56)$.

a Find the value of p and the value of q. **(3 marks)**

b On a grid, sketch the graph of $y = pq^x$ showing any intersections with the axes. **(3 marks)**

14 **Reasoning** Here are the first few rows of Pascal's triangle.

```
            1
          1   1
        1   2   1
      1   3   3   1
    1   4   6   4   1
  1   5  10  10   5   1
```

a Using this information, expand $(4 + x)^3$.

Simplify the expression.

b Given that $(a+b)^5 = a^5 + 5a^{4b} + 10a^{3b^2} + 10a^{2b^3} + 5ab^4 + b^5$, work out the value of

$3^5 + 35 \times 3^4 + 10 \times 3^3 \times 7^2 + 10 \times 3^2 \times 7^3 + 15 \times 7^4 + 7^5$

c Expand and simplify $(2y - 5)^4$.

15 **Problem-solving** The diagram shows the sector, OABC, of a circle.

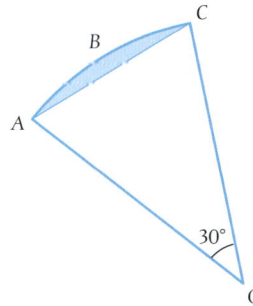

OAC is a triangle where angle AOC = 30°.

Given that the area of the shaded part of the diagram is 7 cm², find the perimeter of the sector.

16 **Problem-solving** Naz has two bags of beads, bag A and bag B.

Bag A contains only 3 red beads, 2 grey beads and 5 blue beads.

Bag B contains only 4 red beads, 1 grey bead and 6 blue beads.

Naz is going to take, at random, a bead from bag A and a bead from bag B.

a Calculate the probability that Naz takes 2 grey beads.

b Calculate the probability that Naz takes 2 beads that are different colours.

Exam-style question

17 a Write the equation $y = 4x^2 + 24x - 5$ in the form $y = a(x + b)^2 + c$, where a, b and c are integers. **(3 marks)**

b Hence, write down the coordinates of the turning point of the equation $y = 4x^2 + 24x - 5$ and state whether it is a maximum or minimum. **(1 mark)**

18 **Problem-solving** The vector $\overrightarrow{BA} = \begin{pmatrix} 2 \\ 3 \end{pmatrix}$ and the vector $\overrightarrow{CA} = \begin{pmatrix} -3 \\ -9 \end{pmatrix}$ are drawn on a centimetre square grid. Find the distance, in cm, between B and C.

19 **Problem-solving** A circle, C, with radius 4 cm and centre (0, 1) is drawn on a grid.

The straight line, L, with equation $y = 2x + 1$ is also drawn on the grid.

Using algebra, find the coordinates of the points of intersection of C and L.

Give the coordinates as exact values in simplified surd form.

Mixed Practice

Exam-style question

20 A bag contains only pink counters and orange counters.

There are 5 more orange counters than pink counters.

A counter is taken at random from the bag and not replaced.

Another counter is taken at random from the bag.

The probability that the two counters are of the same colour is $\frac{77}{155}$.

Work out two different possibilities for the number of orange counters that could be in the bag.

Show clear algebraic working. **(5 marks)**

21 **Problem-solving** The diagram shows the cuboid ABCDEFGH.

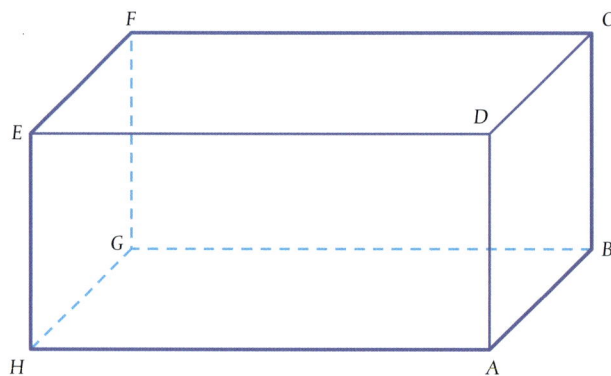

The cuboid is such that ABCD is a square and $AH = 3AB$.

Calculate the size of angle FAG.

Exam-style question

22 Draw a coordinate grid using values of x and y from 0 to 8.

a Show, by shading on the grid, the area defined by

$x \geq 2, y \geq 1, 2x + 3y \leq 15$. **(4 marks)**

b Find the maximum value of $3x + 4y$ in the region where

$x \geq 2, y \geq 1, 2x + 3y \leq 15$. **(2 marks)**

23 **Reasoning** Without using a calculator and showing all the stages of working, write $\dfrac{3 - \sqrt{2}}{(\sqrt{2} - 1)^2}$ in the form $p + q\sqrt{2}$ where p and q are integers.

Mixed Practice

Exam-style question

24 Find, by using algebra, the coordinates of the points of intersection of the straight line with equation $x + y = 7$ and the circle with radius 5 cm and centre $(2, 0)$. **(5 marks)**

25 a Sketch the graph of $y = \cos x$ for $-360° \leq x \leq 360°$

  b On the same axes, sketch the graph of $y = \cos 2x$ for $-360° \leq x \leq 360°$

26 a **Reasoning** Solve the inequality $5x + 3 \leq 7x - 6$.

  b Solve the inequality $2x^2 - 7x < 30$.

  c Write down the set of values for which **both** $5x + 3 \leq 7x - 6$ **and** $2x^2 - 7x < 30$.

Exam-style question

27 Find the value of x such that $16^{-\frac{3}{4}} \times 2^{5x-1} \times 4^{\frac{3}{2}} \times 32^{\frac{2}{5}} = 8$. **(4 marks)**

28 Problem-solving $OABC$ is a quadrilateral.

$\overrightarrow{OA} = 5\mathbf{a}$ $\overrightarrow{OC} = 2\mathbf{b}$ $\overrightarrow{CB} = 4\mathbf{b} - 2\mathbf{a}$

The point P lies on AC and on OB.

  a Find, in terms of **a** and **b**, the vector $\overrightarrow{OB}$

  b Find, in terms of **a** and **b**, the vector $\overrightarrow{AB}$

  Given that $OP = nOB$

  c Using a vector method, find the value of n.

29 Problem-solving The prism $ABCDEFGHIJ$ has a pentagonal end $ABCDE$.

$AE = BC = FG = HI = 3$ cm

$AB = EC = GH = FI = 10$ cm

$AG = EF = CI = BH = 10$ cm

$ED = CD = FJ = IJ = 8$ cm

angle EAB = angle CBA = angle FGH = angle $IHG = 90°$

M is the midpoint of GH.

Find the size of angle AJM.

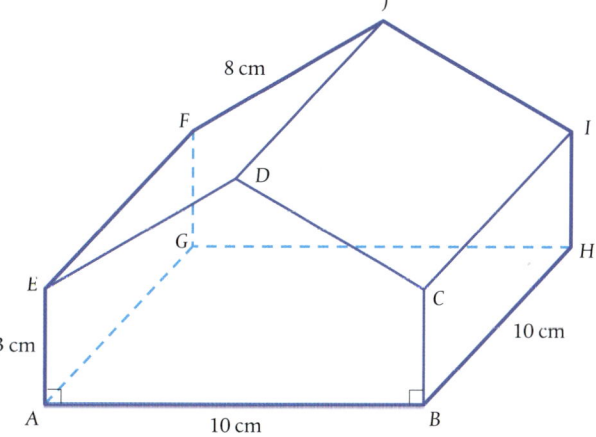

Mixed Practice

Exam-style question

30 Use the trapezium rule to find an estimate for the area of the region under the curve with equation $y = x^2 + 2$ and between the line $x = 1$, the line $x = 5$ and the x-axis.

Use four strips of equal width. **(3 marks)**

31 a Reasoning Use the factor theorem to show that $(x - 3)$ is a factor of

$x^3 - 6x^2 - x + 30$.

b Hence or otherwise, and showing all working, solve

$x^3 - 6x^2 - x + 30 = 0$.

32 Reasoning Write $-5x^2 + 20x + 3$ in the form $a(x + b)^2 + c$, where a, b and c are integers.

33 Reasoning Curve C has equation $y = f(x)$.

The graph of the curve has one minimum point.

The coordinates of this minimum point are $(5, 3)$

Write down the coordinates of the minimum point on the curve with equation

a $y = 4f(x)$ **b** $y = f(0.5x)$

c $y = f(x) - 7$ **d** $y = f(x - 3)$

34 Reasoning Solve $10 \sin x° - 3 = 2$ for $0° \leq x \leq 360°$.

Exam-style question

35 A circle, C, has equation $(x - 3)^2 + (y - 2)^2 = 25$.

Find the equation of the tangent to C at the point where $x = 7$ and y is a positive value. **(5 marks)**

36 Reasoning Solve $2^{2x} - 9 \times 2^x + 8 = 0$

37 Problem-solving There are c counters in a box, $c > 7$.

5 of the counters in the box are pink; the rest of the counters are blue.

Becky is going to take at random 2 counters from the box.

The probability that both counters will be the same colour is $\frac{2}{3}$.

By using algebra and showing all working clearly, find the value of c.

38 Reasoning Given that n is either 1 more or 1 less than a multiple of 7,

prove that n^2 is always 1 more than a multiple of 7.

Mixed Practice

39 Reasoning By drawing a grid for values of x from 0 to 6 and for y from 0 to 12, maximise the value of $3x + 5y$ in the region defined by the inequalities

$y \geq x$, $x \geq 1$, $2x + y \leq 12$.

40 Reasoning A curve A has equation $y = x^2 + 2x + 7$

The translation $\begin{pmatrix} -5 \\ 0 \end{pmatrix}$ transforms curve A to curve B.

Find an equation for B.

Give the answer in the form $y = ax^2 + bx + c$.

41 Reasoning a, b and c are three consecutive multiples of 3.

Prove algebraically that for any set of three consecutive multiples of 3, the difference between the square of the largest multiple of 3 and the smallest multiple of 3 is always a multiple of 18.

Exam-style question

42 $f(x) = 2x^3 - 9x^2 - 2x + 24$

    a Use the factor theorem to show that $(x - 2)$ is a factor of $f(x)$. **(2 marks)**

    b Hence, factorise $f(x)$ fully. **(2 marks)**

43 Reasoning Given that $\sin^2 x + \cos^2 x = 1$ is true for all values of x.

Solve $2 - \sin^2 x + 2\cos x = 0$ for $0° \leq x \leq 360°$.

44 Reasoning The functions f, g and h are defined as

$f : x \Rightarrow 4x^3 + 24x^2 + 27x - 20$

$g : x \Rightarrow \dfrac{x - 4}{5 - 3x}$

$h : x \Rightarrow 3 - \sqrt{x + 4}$

    a Which value of x must be excluded from the domain of x?

    b Find the value of $g(3)$.

    c Find the value of $fg(2)$.

    d Express the inverse function h^{-1} in the form $h^{-1}(x) = ax^2 + bx + c$.

    e Use the factor theorem to show that $(x + 4)$ is a factor of $f(x)$.

    f Hence solve the equation $f(x) = 0$.

      Show clear algebraic working.

45 Reasoning Prove that $4y + 3x = -14$ is a tangent to the circle with equation $x^2 + 10x + y^2 + 12y + 36 = 0$

Answers

Chapter 1: Number

Maths challenge: 60

1.1 Laws of indices

1. a ± 7 b $\pm \frac{1}{9}$ c $\pm \frac{5}{6}$ d 5
   e $\frac{1}{2}$ f $-\frac{3}{4}$
2. a $\pm \frac{1}{5}$ b ± 4 c $\pm \frac{10}{7}$ d $\frac{1}{2}$
   e 4 f $-\frac{10}{7}$
3. a ± 125 b 4 c $\pm \frac{1}{27}$ d $\pm \frac{1}{216}$
   e $-\frac{16}{81}$ f 16
4. a $n=5$ b $n=-3$ c $n=\frac{3}{2}$
5. a 3^{-1} b $3^{\frac{3}{5}}$ c $3^{\frac{16}{3}}$
6. $a=3$
7. $x=-5\frac{2}{5}$
8. $t=3$
9. $x=-18$
10. $y=3$
11. $x=\frac{1}{2}$

1.2 Surds

1. a $2\sqrt{5}$ b $3\sqrt{2}$ c $2\sqrt{11}$ d $3\sqrt{7}$
   e $7\sqrt{2}$ f $10\sqrt{7}$
2. $6\sqrt{5}$
3. a $\sqrt{2}$ b $\sqrt{3}$ c $2\sqrt{3}$
4. a $7\sqrt{3}$ b $4\sqrt{3}$ c $9\sqrt{5}$ d $10\sqrt{2}$
   e $13\sqrt{3}$ f $-8\sqrt{5}$
5. a $2\sqrt{5}-\sqrt{15}$ b $4\sqrt{3}-\sqrt{15}$
   c $7\sqrt{2}+2$ d $20+5\sqrt{2}-4\sqrt{3}-\sqrt{6}$
   e $8+2\sqrt{5}-4\sqrt{3}-\sqrt{15}$ f $11+6\sqrt{3}$
   g $8-5\sqrt{2}$ h $-1+\sqrt{7}$
   i $21-8\sqrt{5}$ j -1
   k $8+2\sqrt{15}$ l $7-2\sqrt{10}$
6. a 72
   b $6\sqrt{5}+30$
7. Ethan has missed some terms out when expanding the brackets.
   $(3+\sqrt{5})^2 = 14+6\sqrt{5}$
8. $(\sqrt{7}-\sqrt{5})(\sqrt{7}+\sqrt{5}) = \sqrt{7}\sqrt{7}+\sqrt{7}\sqrt{5}-\sqrt{5}\sqrt{7}-\sqrt{5}\sqrt{5}$
   $= \sqrt{49}-\sqrt{25}$
   $= 7-5$
   $= 2$

1.3 Rationalising denominators

1. a $\frac{\sqrt{3}}{3}$ b $\frac{\sqrt{7}}{7}$ c $\frac{3\sqrt{2}}{2}$ d $\sqrt{2}$
   e $\sqrt{5}$ f $\frac{\sqrt{6}}{2}$ g $\frac{\sqrt{6}}{3}$ h $\frac{\sqrt{10}}{5}$
   i $\frac{\sqrt{21}}{7}$
2. a $\frac{5-\sqrt{3}}{22}$ b $\frac{3+\sqrt{2}}{7}$ c $\frac{1+\sqrt{5}}{-2}$ d $-2+2\sqrt{2}$
   e $6+3\sqrt{3}$ f $\frac{10-2\sqrt{7}}{9}$ g $\frac{\sqrt{2}-\sqrt{6}}{-2}$ h $\frac{3+\sqrt{3}}{-2}$
   i $\frac{2\sqrt{3}+\sqrt{6}}{2}$ j $\frac{3\sqrt{5}-\sqrt{15}}{6}$ k $\frac{4\sqrt{5}+5}{11}$ l $\frac{1+\sqrt{7}}{6}$
3. a $-\sqrt{2}-\sqrt{3}$ b $\frac{5}{2}(-\sqrt{3}+\sqrt{5})$
   c $\frac{3+\sqrt{21}}{-2}$ d $\frac{2\sqrt{3}-2\sqrt{5}+3-\sqrt{15}}{-2}$
   e $-5-2\sqrt{6}$ f $\frac{\sqrt{77}-9}{2}$
4. a $7-4\sqrt{3}$ b $9+4\sqrt{5}$ c $\frac{42+15\sqrt{3}}{242}$ d $\frac{38-12\sqrt{2}}{289}$
   e $\frac{4-\sqrt{2}}{14}$ f $\frac{-9+6\sqrt{5}}{11}$ g $\frac{2\sqrt{3}-3}{2}$ h $\frac{27\sqrt{5}+10\sqrt{10}}{529}$
   i $\frac{8-3\sqrt{2}}{23}$
5. a $3\sqrt{3}$ b $\frac{4\sqrt{3}+\sqrt{15}}{9}$
6. a $\frac{2\sqrt{2}+\sqrt{10}}{6}$ b $\frac{\sqrt{3}-1}{8}$
   c $\frac{\sqrt{5}+\sqrt{10}}{35}$ d $\frac{\sqrt{6}-4\sqrt{2}}{10}$
   e $\frac{2+12\sqrt{3}+\sqrt{5}+6\sqrt{15}}{-107}$ f $\frac{5+5\sqrt{5}+\sqrt{3}+\sqrt{15}}{-8}$
7. $\frac{4+\sqrt{7}}{3}$
8. $\frac{2+\sqrt{3}}{\sqrt{80}-\sqrt{20}} = \frac{2+\sqrt{3}}{4\sqrt{5}-2\sqrt{5}}$
   $= \frac{2+\sqrt{3}}{2\sqrt{5}}$
   $= \frac{(2+\sqrt{3})\times\sqrt{5}}{2\sqrt{5}\times\sqrt{5}}$
   $= \frac{2\sqrt{5}+\sqrt{15}}{10}$

Chapter 2: Algebraic manipulation

Maths challenge: The book must have 450 pages so 86 5s are used

2.1 Algebraic indices

1. a x^9 b $12a^9$ c $3p^3$
   d $7q^3$ e $5y^3$ f $9b^6$
   g $3z^{14}$ h $80m^{14}$ i $72t^{17}$
   j $8n^5$ k $25k^8$ l $8c^{11}$
   m $500v^9$ n $144x^{15}$ o $16y^{-2}$
2. a x^4+x^7 b $4t^4+7t^2$ c $2y^3-\frac{5}{2}$
   d $\frac{4b}{3}+b^6$ e $2-\frac{2z}{3}$ f $\frac{3n^3}{5}-n$
3. a $a=\frac{1}{81}$ and $n=-8$.
   b $c=12$, $d=10$ and $n=1$

Answers

4 a $y^{\frac{3}{2}}$ **b** $8a^4$ **c** $2x^{\frac{3}{2}}$
  d $4b^{\frac{1}{2}}$ **e** $6t^{\frac{5}{2}}$ **f** $\pm 5m^{\frac{3}{2}}$

5 x^2

6 a $\frac{x}{3}$ **b** $81x^{-5}$

7 LHS $= \dfrac{(4-\sqrt{x})^2}{\sqrt{x}}$

$= \dfrac{\left(4-x^{\frac{1}{2}}\right)\left(4-x^{\frac{1}{2}}\right)}{x^{\frac{1}{2}}}$

$= \dfrac{16 - 8x^{\frac{1}{2}} + x}{x^{\frac{1}{2}}}$

$= \dfrac{16}{x^{\frac{1}{2}}} - \dfrac{8x^{\frac{1}{2}}}{x^{\frac{1}{2}}} + \dfrac{x}{x^{\frac{1}{2}}}$

$= 16x^{-\frac{1}{2}} - 8 + x^{\frac{1}{2}} = $ RHS

8 $a = 3$ and $b = \dfrac{7}{2}$ or 3.5

2.2 Expanding brackets

1 a $x^2 + x - 12$ **b** $x^2 - 15x + 56$
  c $x^2 - 12x + 36$ **d** $x^2 - 6xy + 9y^2$
  e $x^2 + 4xy - 45y^2$ **f** $2x^2 - 19xy + 24y^2$
  g $25x^2 - 40xy + 16y^2$ **h** $x^3 - 3x + 2x^2y - 6y$
  i $x^3 + 4x^2 + 5x + 2$ **j** $2x^2 + 4xy + x + 6y - 3$
  k $9xy - 6y - 12x^2 + 11x - 2$ **l** $x^3 - 4xy^2 + 3x^2 - 12y^2$
  m $2x^3 - 5x^2 - 18x - 35$ **n** $3x^3 - 7x^2 - 2x + 8$
  o $5x^3 + 19x^2 - 6x - 8$

2 a $a^2 + 2ab + b^2$ **b** $4y^2 + 12yz + 9z^2$

3 a $x^3 + 5x^2 + 6x$ **b** $x^3 - 3x^2 - 4x$
  c $2x^3 + 7x^2 - 4x$ **d** $6x^3 - 17x^2 + 12x$
  e $8x^3 - 2x^2 - 6x$ **f** $30x^3 - 25x^2 - 20x$
  g $x^3 - x^2y + 4x^2 - xy + 3x$ **h** $6x^3 + 9x^2y - 15x^2 - 18xy + 6x$
  i $2x^3 + 4x^2 + 5x^2y - 3xy^2 + 12xy$ **j** $x^3 + 6x^2 + 11x + 6$
  k $x^3 + 4x^2 + x - 6$ **l** $x^3 + 5x^2 + 2x - 8$
  m $x^3 + x^2 - 10x + 8$ **n** $x^3 - 8x^2 + 19x - 12$
  o $x^3 - 4x^2 - 11x + 30$ **p** $2x^3 - x^2 - 25x - 12$
  q $6x^3 - 23x^2 + 25x - 6$ **r** $15x^3 + 61x^2 + 2x - 8$
  s $x^3 + 3x^2 + 2x + x^2y + 3xy + 2y$ **t** $x^3 - 7x^2 + 2x^2y - 14xy + xy^2 - 7y^2$
  u $27x^3 + 54x^2y + 36xy^2 + 8y^3$

4 a $c^3 + 3c^2d + 3cd^2 + d^3$
  b $a^4 + 4a^3b + 6a^2b^2 + 4ab^3 + b^4$
  c $p^5 + 5p^4q + 10p^3q^2 + 10p^2q^3 + 5pq^4 + q^5$

5 a i $x^3 + 6x^2y + 12xy^2 + 8y^3$
  ii $x^3 - 6x^2y + 12xy^2 - 8y^3$
  iii $8x^3 - 36x^2y + 54xy^2 - 27y^3$
  b i $x^4 + 8x^3y + 24x^2y^2 + 32xy^3 + 16y^4$
  ii $x^4 - 8x^3y + 24x^2y^2 - 32xy^3 + 16y^4$
  iii $16x^4 - 96x^3y + 216x^2y^2 - 216xy^3 + 81y^4$

6 a i $x^3 + 6x^2 + 12x + 8$
  ii $x^3 - 6x^2 + 12x - 8$
  iii $8x^3 - 36x^2 + 54x - 27$
  b i $x^4 + 8x^3 + 24x^2 + 32x + 16$
  ii $x^4 - 8x^3 + 24x^2 - 32x + 16$
  iii $16x^4 - 96x^3 + 216x^2 - 216x + 81$
  c $32x^5 - 80x^4 + 80x^3 - 40x^2 + 10x - 1$

7 a $1 + 40x + 600x^2 + 4000x^3 + 10\,000x^4$
  b $1\,004\,006\,004\,001$

8 $p^4 + 4p^3q + 6p^2q^2 + 4pq^3 + q^4$

9 $1 + 2x + \dfrac{3x^2}{2} + \dfrac{x^3}{2} + \dfrac{x^4}{16}$

10 $8a^3 + 12a^2b + 6ab^2 + b^3$

11 a $2x^4 + 12x^2y^2 + 2y^4$
  b $33x^5 + 70x^4y + 120x^3y^2 - 40x^2y^3 + 90xy^4 - 31y^5$

2.3 Factorising

1 a $5(2x - 5)$ **b** $6(3x^2 + 2)$
  c $\dfrac{1}{2}x(x - 3)$ **d** $(x + 4)(x - 4)$
  e $\dfrac{2}{5}x(x + 6)$ **f** $\dfrac{2}{7}x(3x + 5)$
  g $\dfrac{4}{3}x(4 - 5x)$ **h** $\dfrac{4}{11}x(2 - 3y)$
  i $(5 + x)(5 - x)$ **j** $(x + 5y)(x - 5y)$
  k $\left(\dfrac{9}{10}x + \dfrac{2}{7}y\right)\left(\dfrac{9}{10}x - \dfrac{2}{7}y\right)$ **l** $4(3 + y)(3 - y)$
  m $(x + 4)(x - 3)$ **n** $(x - 1)(x + 5)$
  o $(x + 5)(x - 4)$ **p** $(x + 6)(x - 5)$
  q $(x - 2)(x - 7)$ **r** $(x + 3)(x - 7)$
  s $(4x + 1)(x - 2)$ **t** $(3x - 2)(x + 3)$
  u $(2x + 1)(x + 4)$ **v** $(4x - 1)(3x - 2)$
  w $(5x - 2)(2x - 1)$ **x** $(3x - 1)(-2x + 3)$

2 a $x^2(x + 3)$ **b** $x^2(2 - 7x)$
  c $4x^2(3 - 5x)$ **d** $x(x + 2)(x + 3)$
  e $x(x - 1)(x + 2)$ **f** $x(x + 3)(x - 3)$
  g $x(x + 5)(x - 4)$ **h** $x(x - 3)(x - 4)$
  i $x(x - 6)(3x + 4)$ **j** $2x(x + 1)(2x + 3)$
  k $3x(2x + 1)(2x - 1)$ **l** $-2x(3x + 1)(2x + 5)$

3 a $1^3 - 1^2 - 10 \times 1 + 10 = 1 - 1 - 10 + 10 = 0$
  b $(-1)^3 + 2 \times (-1)^2 - 6 \times (-1) - 7 = -1 + 2 + 6 - 7 = 0$
  c $(-2)^3 + (-2)^2 - 3 \times (-2) - 2 = -8 + 4 + 6 - 2 = 0$
  d $3^3 - 2 \times 3^2 - 5 \times 3 + 6 = 27 - 18 - 15 + 6 = 0$
  e $5^3 - 3 \times 5^2 - 7 \times 5 - 15 = 125 - 75 - 35 - 15 = 0$
  f $(-4)^3 + 5 \times (-4)^2 - (-4) - 20 = -64 + 80 + 4 - 20 = 0$

4 a
$$\begin{array}{r}x^2 - 10\\x-1{\overline{\smash{\big)}\,x^3 - x^2 - 10x + 10}}\\\underline{x^3 - x^2}\\0 \quad -10x + 10\\\underline{-10x + 10}\\0\end{array}$$

177

Answers

b $x+1 \overline{\smash{)}\begin{array}{l} x^2 + x - 7 \\ x^3 + 2x^2 - 6x - 7 \end{array}}$
$\underline{x^3 + x^2}$
$ x^2 - 6x$
$ \underline{x^2 + x}$
$ -7x - 7$
$ \underline{-7x - 7}$
$ 0$

c $x+2 \overline{\smash{)}\begin{array}{l} x^2 - x - 1 \\ x^3 + x^2 - 3x - 2 \end{array}}$
$\underline{x^3 + 2x^2}$
$ -x^2 - 3x$
$ \underline{-x^2 - 2x}$
$ -x - 2$
$ \underline{-x - 2}$
$ 0$

d $x-3 \overline{\smash{)}\begin{array}{l} x^2 + x - 2 \\ x^3 - 2x^2 - 5x + 6 \end{array}}$
$\underline{x^3 - 3x^2}$
$ x^2 - 5x$
$ \underline{x^2 - 3x}$
$ -2x + 6$
$ \underline{-2x + 6}$
$ 0$

e $x-5 \overline{\smash{)}\begin{array}{l} x^2 + 2x + 3 \\ x^3 - 3x^2 - 7x - 15 \end{array}}$
$\underline{x^3 - 5x^2}$
$ 2x^2 - 7x$
$ \underline{2x^2 - 10x}$
$ 3x - 15$
$ \underline{3x - 15}$
$ 0$

f $x+4 \overline{\smash{)}\begin{array}{l} x^2 + x - 5 \\ 4x^3 + 5x^2 - x - 20 \end{array}}$
$\underline{x^3 + 4x^2}$
$ x^2 - x$
$ \underline{x^2 + 4x}$
$ -5x - 20$
$ \underline{-5x - 20}$
$ 0$

5 a $2 \times (-3)^3 + 9 \times (-3)^2 + 4 \times (-3) - 15 = -54 + 81 - 12 - 15 = 0$
 b $(x + 3)(x - 1)(2x + 5)$

6 a $(-5)^3 + 4 \times (-5)^2 - 25 \times (-5) - 100 = -125 + 100 + 125 - 100 = 0$
 b $(x + 4)(x + 5)(x - 5)$

7 a $1^3 + 3 \times 1^2 - 4 = 1 + 3 - 4 = 0$
 b $(x - 1)(x + 2)^2$

8 $(x + 3)(2x + 1)(3x - 2)$

9 $(x + 4)^2(2x + 5)$

10 a $5 \times 4^3 - 17 \times 4^2 - 5 \times 4 - 28 = 320 - 272 - 20 - 28 = 0$
 b $5x^3 - 17x^2 - 5x - 28 = (x - 4)(5x^2 + 3x + 7)$

11 a $x+3 \overline{\smash{)}\begin{array}{l} 4x^2 - 12x + 9 \\ 4x^3 + 0x^2 - 27x + 27 \end{array}}$
$\underline{-(4x^3 + 12x^2)}$
$ -12x^2 - 27x$
$ \underline{-(-12x^2 - 36x)}$
$ 9x + 27$
$ \underline{-(9x + 27)}$
$ 0$

b $(x + 3)(2x - 3)^2$

12 $(x - 2)(x + 1)(x - 1)$

2.4 Completing the square

1 a $(x + 6)^2$ **b** $(x - 5)^2$
c $\left(x + \frac{1}{2}\right)^2$

2 a $(x + 6)^2 - 36$ **b** $(x - 7)^2 - 49$
c $\left(x + \frac{9}{2}\right)^2 - \frac{81}{4}$

3 a $\left(x + \frac{5}{2}\right)^2 - \frac{41}{4}$ **b** $\left(x - \frac{11}{2}\right)^2 - \frac{133}{4}$
c $\left(x - \frac{1}{2}\right)^2 + \frac{7}{4}$

4 a $2\left(x - \frac{1}{4}\right)^2 - \frac{25}{8}$ **b** $3\left(x + \frac{5}{6}\right)^2 - \frac{49}{12}$
c $5\left(x - \frac{2}{5}\right)^2 + \frac{1}{5}$ **d** $4\left(x + \frac{1}{4}\right)^2 - \frac{13}{4}$
e $2\left(x + \frac{7}{4}\right)^2 + \frac{23}{8}$ **f** $3\left(x - \frac{1}{6}\right)^2 - \frac{61}{12}$

5 $5\left(x - \frac{1}{5}\right)^2 - \frac{6}{5}$

6 $A = 3, B = \frac{7}{6}$ and $C = -\frac{97}{12}$

7 $\left(3x + \frac{5}{2}\right)^2 + \frac{15}{4}$

2.5 Algebraic fractions

1 a $\frac{x}{2}$ **b** 3
c $\frac{1}{x + 4}$ **d** $\frac{1}{x - 4}$
e $\frac{x - 3}{x + 2}$ **f** $\frac{x + 1}{x - 3}$
g $\frac{x - 3}{x - 7}$ **h** $\frac{2x + 1}{x + 3}$
i $\frac{3x}{x + 5}$ **j** $\frac{2x + 5}{2x - 3}$
k $\frac{2x - 5}{2x - 1}$ **l** $\frac{x(x + 4)}{x + 5}$
m $\frac{x(2x + 1)}{x + 3}$ **n** $\frac{x(2x + 5)}{3x + 1}$
o $\frac{3(x + 4)}{x + 3}$

2 $\frac{x + 3}{x + 2}$

3 a $(-1)^3 - (-1)^2 - 17 \times (-1) - 15 = -1 - 1 + 17 - 15 = 0$
 b $(x + 1)(x + 3)(x - 5)$
 c $\frac{x + 1}{2x}$

4 a i $2 \times 3^3 + 3^2 - 15 \times 3 - 18 = 54 + 9 - 45 - 18 = 0$
 ii $3^3 - 2 \times 3^2 - 5 \times 3 + 6 = 27 - 18 - 15 + 6 = 0$
 b i $(x - 3)(2x + 3)(x + 2)$
 ii $(x - 3)(x + 2)(x - 1)$
 c $A = 2, B = 3$ and $C = -1$

Answers

5 a $\frac{3}{5}$ b $\frac{4(x+1)}{5x}$

   c $\frac{3}{4}$ d $\frac{1}{6}$

   e 6   f $\frac{x}{x+3}$

   g $\frac{x-7}{x}$ h $\frac{2x+3}{2x-1}$

   i $\frac{4x-3}{2x+5}$

6 a $(-2)^3 - 19 \times (-2) - 30 = -8 + 38 - 30 = 0$

   b $(x+2)(x+3)(x-5)$

   c $\frac{1}{x+3}$

7 a $2 \times \left(\frac{1}{2}\right)^3 + \left(\frac{1}{2}\right)^2 - 25 \times \left(\frac{1}{2}\right) + 12 = \frac{1}{4} + \frac{1}{4} - \frac{25}{2} + 12 = 0$

   b $(2x-1)(x-3)(x+4)$

   c $\frac{x+5}{3x}$

8 a $\frac{x}{6}$ b $\frac{5x}{12}$

   c $\frac{11x}{15}$ d $\frac{5}{3x}$

   e $\frac{11}{2x}$ f $-\frac{1}{15x}$

   g $\frac{5x+11}{6}$ h $\frac{x+9}{12}$

   i $\frac{5x+9}{4}$ j $\frac{3x+10}{(x+3)(x+4)}$

   k $\frac{2(x-7)}{(x+1)(x-3)}$ l $\frac{2x-11}{(x-2)(x-3)}$

   m $\frac{2x+13}{x+5}$ n $\frac{x-1}{x+4}$

   o $\frac{2x^2+5x+1}{(x-1)(x+3)}$ p $\frac{2x+2}{(x+3)(x+4)}$

   q $\frac{-7x-4}{(x+4)(x-2)}$ r $\frac{2x-3}{x-5}$

9 $\frac{2x^2+10}{(x-1)(x+1)}$

10 a $\frac{x+1}{x(x+4)}$ b $\frac{3x-1}{x(x+2)}$

   c $\frac{-x}{(x+1)(x+2)}$ d $\frac{2}{x-1}$

   e $\frac{x-3}{x+2}$ f $\frac{-x-15}{(x-1)(x-5)}$

   g $\frac{3x^2+10x+7}{(x+3)(x-1)}$ h $\frac{-3x-1}{(x+1)(x-1)}$

   i $\frac{5x^2-14x+8}{(x-5)(x+4)}$

11 $\frac{3x^3+17x^2+10x}{2x^3-3x^2+x} \div \frac{4x^3+23x^2+15x}{2x^3+7x^2-4x}$

$= \frac{3x^3+17x^2+10x}{2x^3-3x^2+x} \times \frac{2x^3+7x^2-4x}{4x^3+23x^2+15x}$

$= \frac{x(3x+2)(x+5)}{x(2x-1)(x-1)} \times \frac{x(2x-1)(x+4)}{x(4x+3)(x+5)}$

$= \frac{3x+2}{x-1} \times \frac{x+4}{4x+3}$

$= \frac{3x^2+14x+8}{4x^2-x-3}$

12 $\frac{2x^2-9x-5}{3x^2-11x-20} + \frac{4x^3-4x^2-3x}{4x^3-x}$

$= \frac{(2x+1)(x-5)}{(3x+4)(x-5)} + \frac{x(2x+1)(2x-3)}{x(2x+1)(2x-1)}$

$= \frac{2x+1}{3x+4} + \frac{2x-3}{2x-1}$

$= \frac{(2x+1)(2x-1)}{(3x+4)(2x-1)} + \frac{(2x-3)(3x+4)}{(2x-1)(3x+4)}$

$= \frac{4x^2-1}{(3x+4)(2x-1)} + \frac{6x^2-x-12}{(2x-1)(3x+4)}$

$= \frac{10x^2-x-13}{(2x-1)(3x+4)}$

13 $\frac{1}{(x-6)(x-3)(x-1)}$ or $\frac{1}{(x-6)(x^2-4x+3)}$

Chapter 3: Graphs

Maths challenge: $\frac{1}{a} + \frac{1}{b} = \frac{b+a}{ab}$

$= \frac{4}{8}$

$= \frac{1}{2}$

3.1 Linear graphs

1 $3x + 2y = 7$, $(y-2) = 5(x-1)$ and $x + 3y - 7 = 0$ pass through (1, 2)

2 a $(y-4) = \frac{1}{2}(x-3)$

   b $2x - 3y + 1 = 0$

   c $4x + 5y - 7 = 0$

   d $y - 7 = -2(x-2)$

3 $y - 3 = 4(x-2)$ and $y = 4x + 3$

4 $y + 3 = 2(x+5)$, $y + 4x - 7 = 0$ and $3x + 2y - 14 = 0$

5 a Gradient $= 2$, $y - -1 = 2(x-4)$

     $y + 1 = 2(x-4)$

   b $y - 6 = -4(x-2)$

   c $y + 2 = \frac{2}{3}(x-3)$

6 a $-\frac{1}{2}$

   b 3

   c $\frac{3}{4}$

   d 2.5

7 a $y = \frac{1}{2}x + 2$

   b $x + 4y - 24 = 0$

   c $(y + 10) = \frac{3}{4}(x - \sqrt{2})$

8 $(y - 1) = 2(x + 2)$ or $(y - 11) = 2(x - 3)$

9 $x + 3y = 6$

10 For example, line PR has gradient $\frac{11-2}{6-3} = \frac{9}{3} = 3$

PR has equation $(y-2) = 3(x-3)$

Substituting the coordinates of Q(4, 5): $(5-2) = 3(4-3)$, which is true.

So, P, Q and R can be joined by the straight line $(y-2) = 3(x-3)$.

179

Answers

11 a

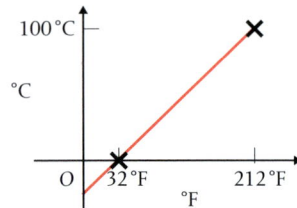

b $y = \frac{5}{9}(x - 32)$ or equivalent

c an increase of 1°F = an increase of $\frac{5}{9}$°C

d The y-intercept is −17.8 (1 d.p.), which is the temperature in °C equivalent to 0°F

12 For example, AB has equation $y = \frac{3}{4}(x + 4)$

BC has equation $y - 3 = 2x$

AC has equation $y = -\frac{1}{2}(x + 4)$

AC and BC are perpendicular, because the product of their gradients is −1

ABC is a right-angled triangle.

3.2 Quadratic graphs

1

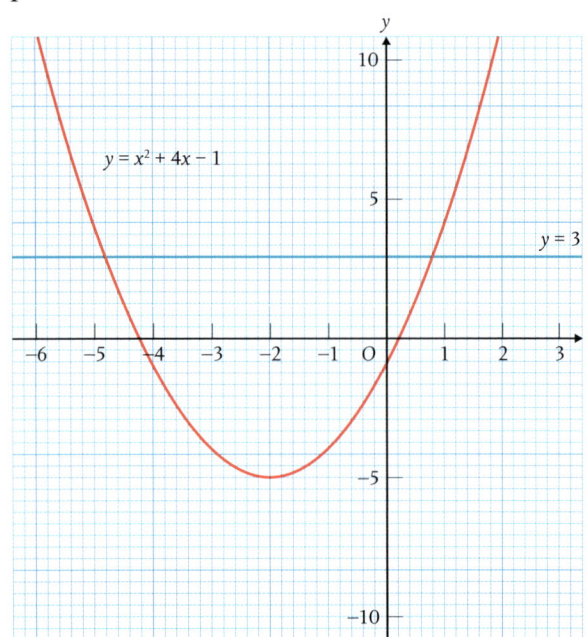

$x = -4.8$ and $x = 0.8$

2

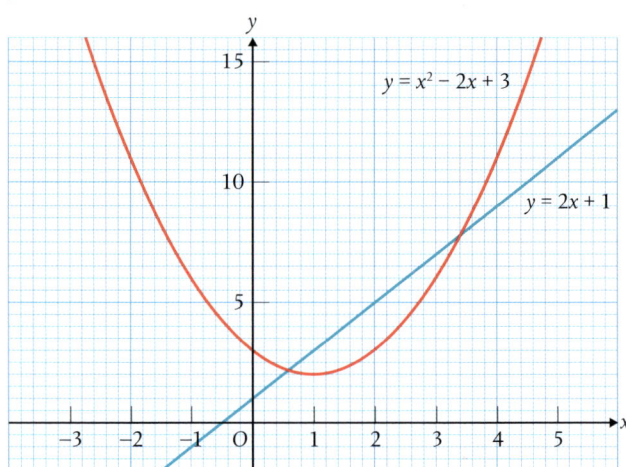

$x = 0.6$ and $x = 3.4$

3 a $x = 4.44, x = 0.56$ (2 d.p.)   **b** $x = -0.63, x = 2.63$ (2 d.p.)

c $x = -0.84, x = 0.24$ (2 d.p.)

4 a $x = \frac{-7 + \sqrt{41}}{4}, x = -\frac{7 + \sqrt{41}}{4}$ **b** $x = \frac{-4 + \sqrt{6}}{2}, x = -\frac{4 + \sqrt{6}}{2}$

c $x = \frac{-1 + \sqrt{7}}{3}, x = -\frac{1 + \sqrt{7}}{3}$

5 a 2   **b** 0   **c** 2

   d 2   **e** 0   **f** 1

6 a (−2, 1) minimum   **b** $\left(-\frac{3}{4}, -\frac{17}{8}\right)$ minimum

   c (2, 3) maximum   **d** (1, 10) maximum

7 a $x = -2.9, x = 0.9$ **b** (−1, −7) minimum

   c $y = -5$

   d

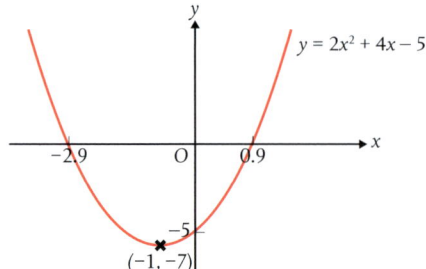

8

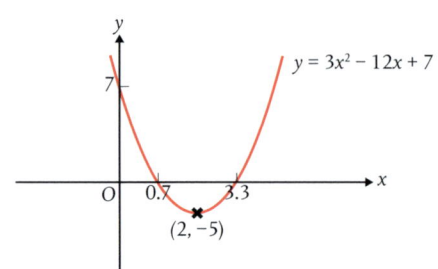

9

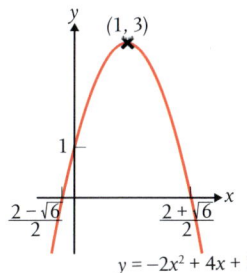

10

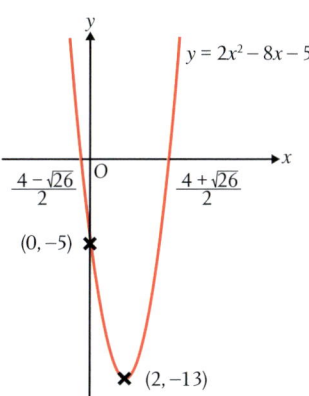

11 a $y = x^2 + 4x - 3$ **b** $-2 + \sqrt{7}, -2 - \sqrt{7}$

12 The graph of $y = 2x^2 - 5x + 6$ has no real roots, because its discriminant is negative.

13 A: $y = 3x^2 + 4x + 9$ **B:** $y = -x^2 + 9$
   C: $y = x^2 - 6x + 9$ **D:** $y = x^2 + 7x + 9$

14 $p = \dfrac{25}{8} = 3\dfrac{1}{8}$ or 3.125

3.3 Cubic and quartic graphs

1

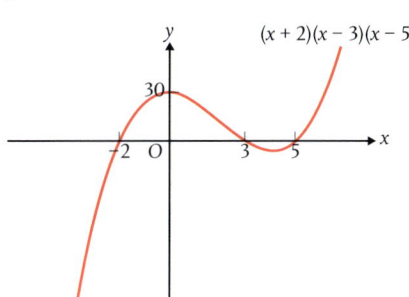

2

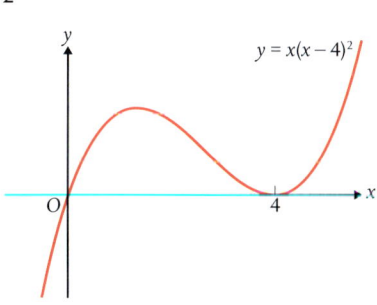

3

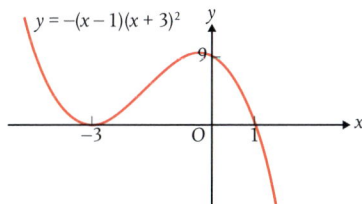

4

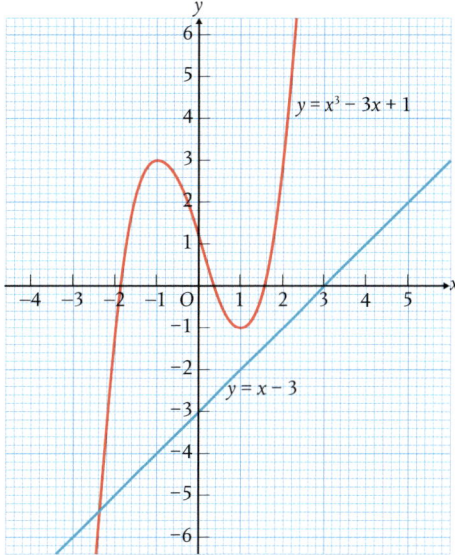

$x = -2.4$

5

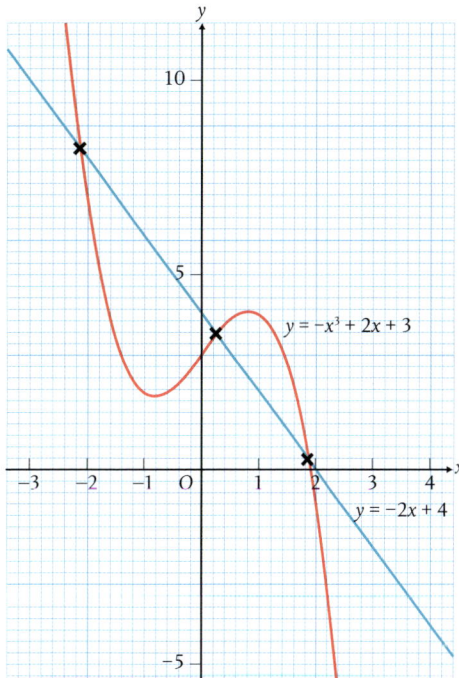

$x = -2.1, x = 0.2$ or $0.3, x = 1.9$ (accept answers ± 0.1)

Answers

6 a $x^3 + 15x^2 + 75x + 125$

b

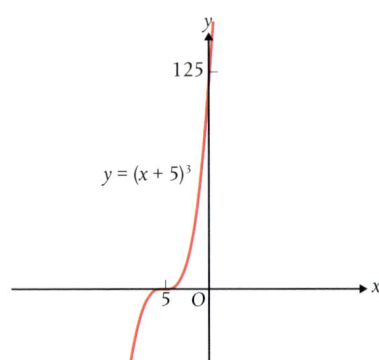

7 a $y = x^3 - 10x^2 + 28x - 24$

b $y = x^3 - 9x^2 + 27x - 27$

8

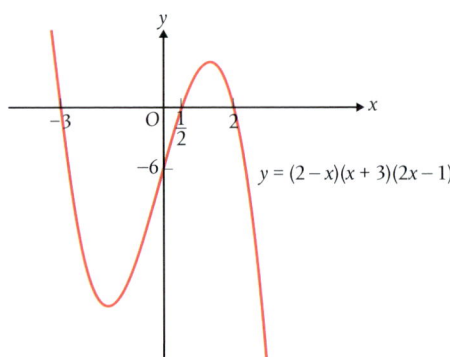

9 a When $x = 1$, $x^3 + 6x^2 + 3x - 10 = 1^3 + 6 + 3 - 10 = 0$

So $(x - 1)$ is a factor of $x^3 + 6x^2 + 3x - 10$

b $x^3 + 6x^2 + 3x - 10 = (x - 1)(x^2 + 7x + 10)$

c $(x - 1)(x^2 + 7x + 10) = (x - 1)(x + 2)(x + 5)$

d

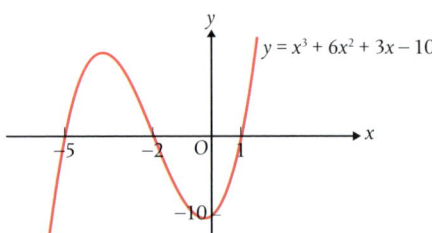

10

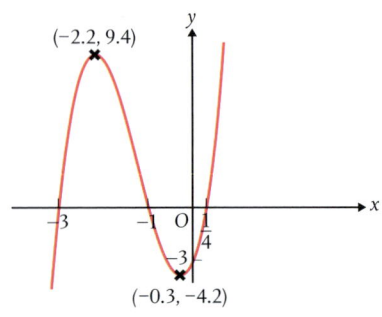

11 $(-2, 0)$ and $(-1, 0)$

12 a

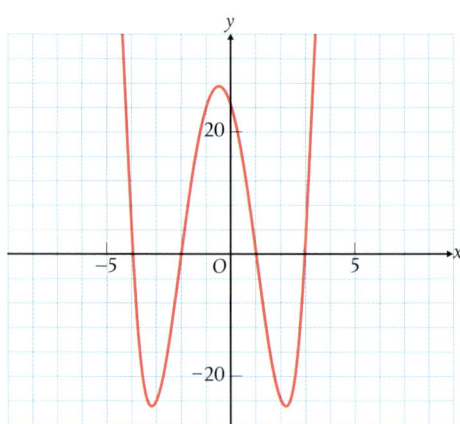

b $(x - 3)(x + 2)(x - 1)(x + 4)$

13 a

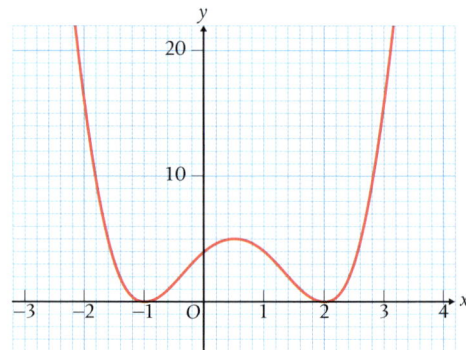

b $(x + 1)^2 (x - 2)$

14

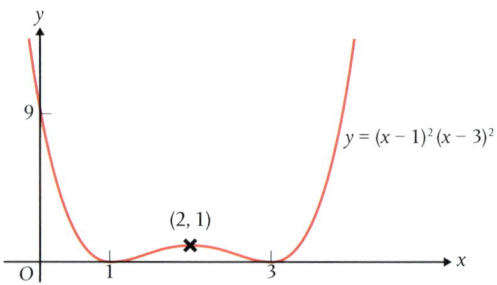

15

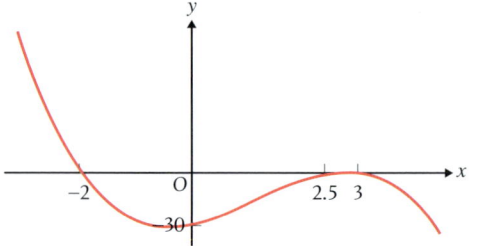

Answers

16

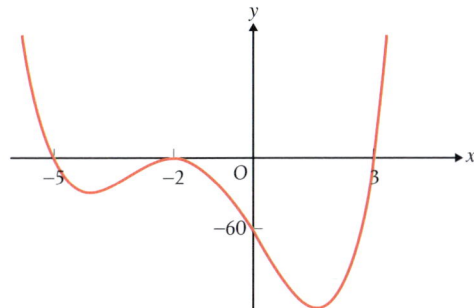

3.4 Trigonometric graphs
1

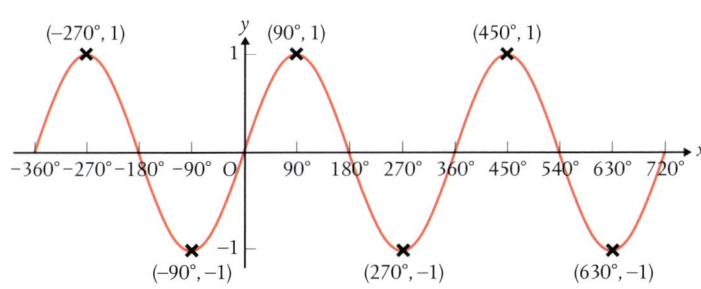

2

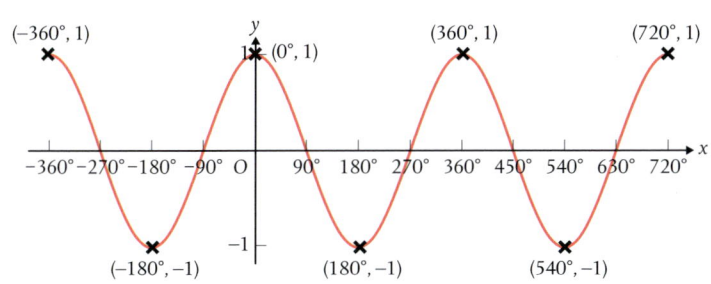

3 a $\tan 30° = \frac{\sqrt{3}}{3}$ b $\sin 45° = \frac{\sqrt{2}}{2}$

4 a i P(360, 1)   ii Q(180, −1)

   b $y = \frac{1}{2}$

   c 60°, 300°, 420°

5 $x = −315°, −135°, 45°, 225°$

6 $x = −315°, −225°, 45°, 135°$

7 a $\cos(−30)° = \frac{\sqrt{3}}{2}$

   b $\cos 300° = 0.5$

8 a $\sin −30° = −0.5$

   b $\sin 210° = −0.5$

9 a 7

   b 3

   c 6

10 Any three negative products of 90 and an odd number, e.g. −90°, −270°, −450°
    Any three positive products of 90 and an odd number, e.g. 90°, 270°, 450°

11 56°, 124°, 416°, 484°, −236°, −304°

12 $x = −330°, −30°, 30°, 330°$

13 $x = 240°$ or $300°$
    $\cos 240° = −0.5$, $\cos 300° = 0.5$

14 −330°, −210°, 30°, 150°

15 −315°, −45°, 45°, 315°

Chapter 4: More graphs

Maths challenge: @ = ×, # = −, ? = 5

4.1 Translating and reflecting graphs

1 a–c

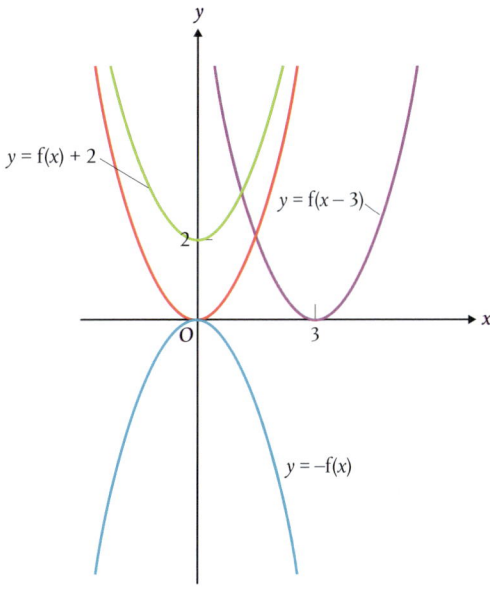

2 a–c

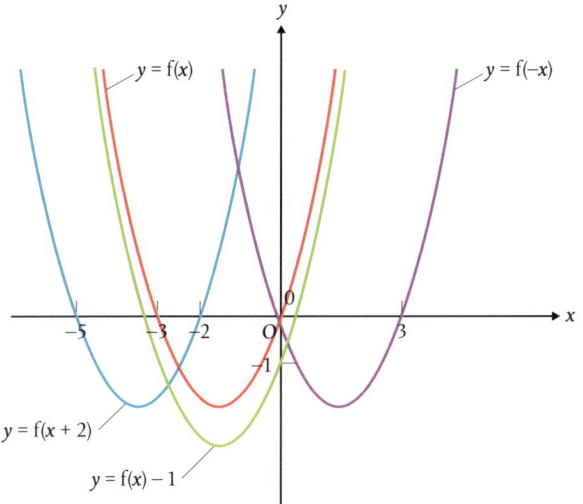

3 a–c

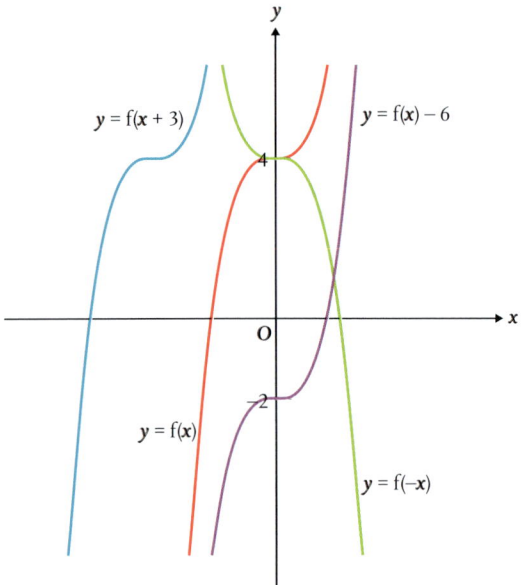

4 a–c

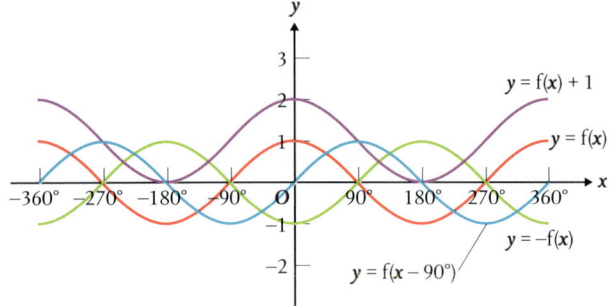

5 a–c

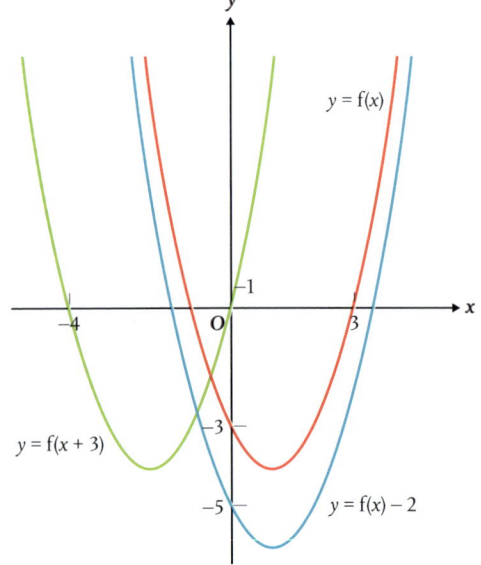

6 a–c

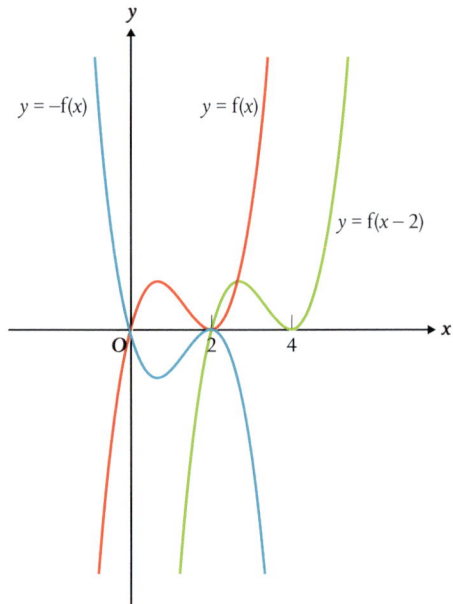

7 a–c

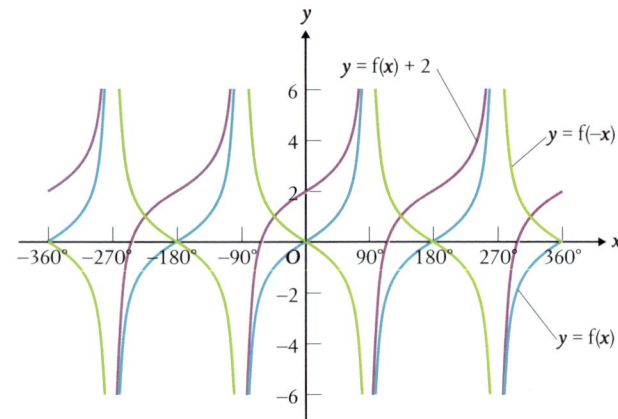

8 a 18

b $y = x^3 - 2x^2 - 5x + 6$

c

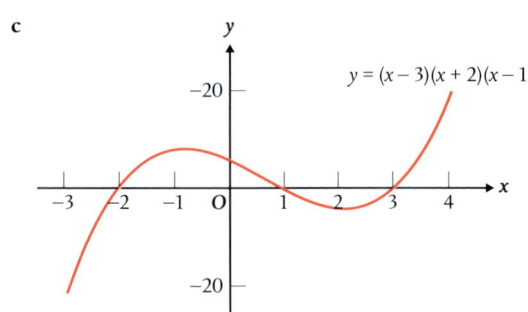

d

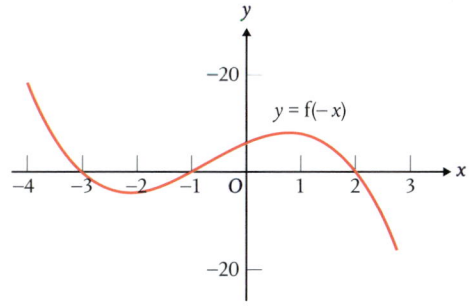

9 a–b

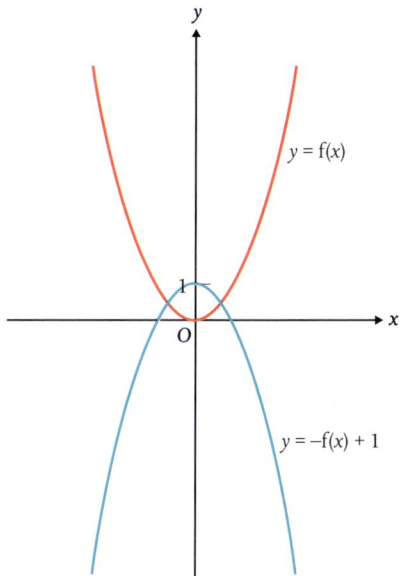

10 a–c

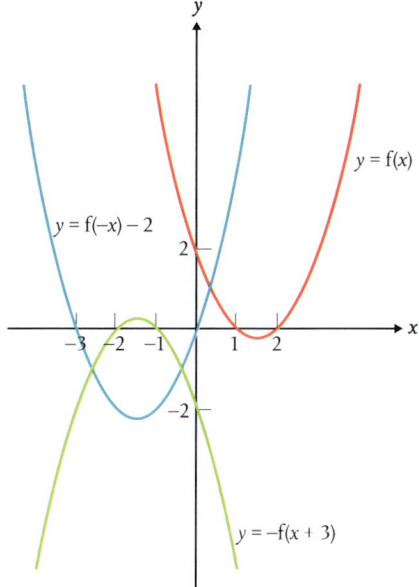

11 a–c

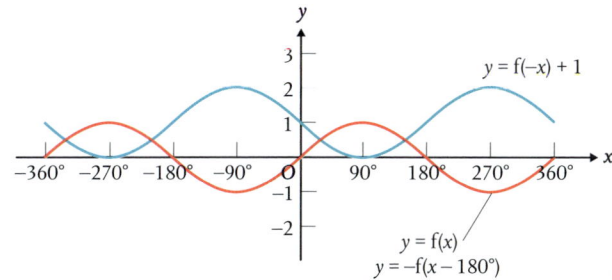

12 a

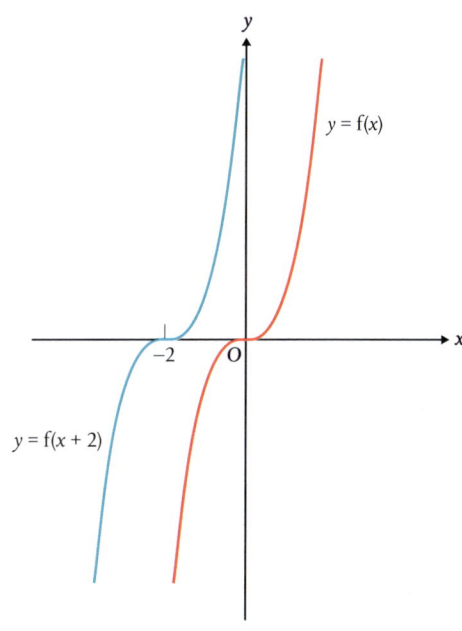

b $(-2, 0)$

13 a 0 and 3

b -3 and 0

c -1 and 2

14 a 0 and 2

b 1 and 3

c -3 and -1

15 a $P(-270°, -2)$, $Q(-90°, -4)$ and $R(180°, -3)$

b $P(-270°, 0)$, $Q(-90°, 2)$ and $R(180°, 1)$

c $P(-180°, 3)$, $Q(0°, 1)$ and $R(270°, 2)$

16 Graph A: $y = f(x) - 3$

Graph B: $y = f(x - 90°) + 2$

Answers

4.2 Stretching graphs

1 a–b

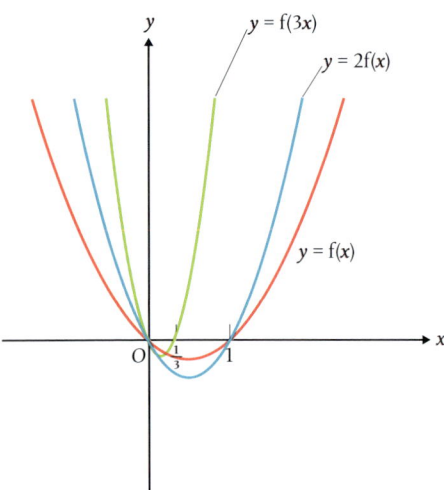

2 a–b

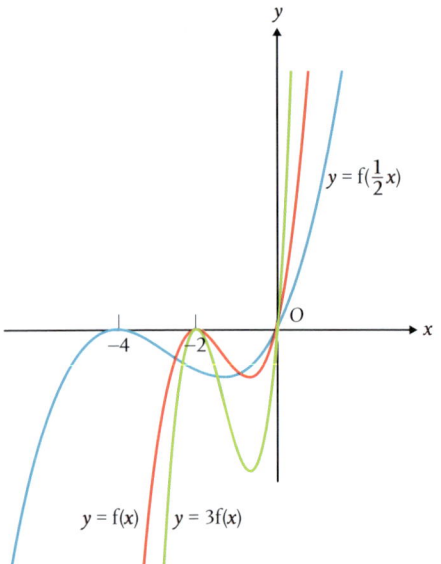

3 a–c

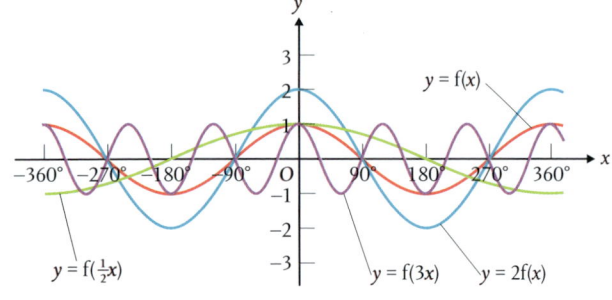

4 a–b

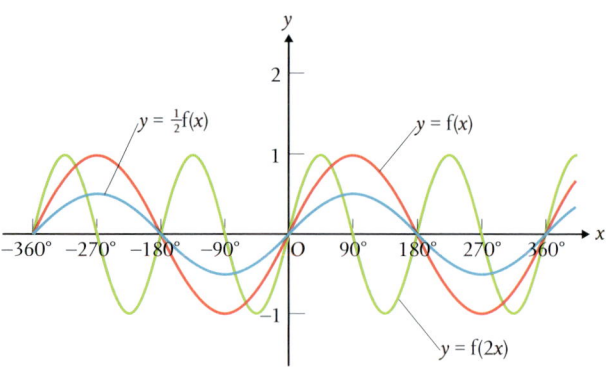

5 a–c

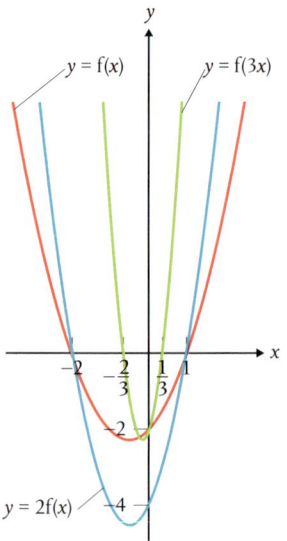

6 a–c

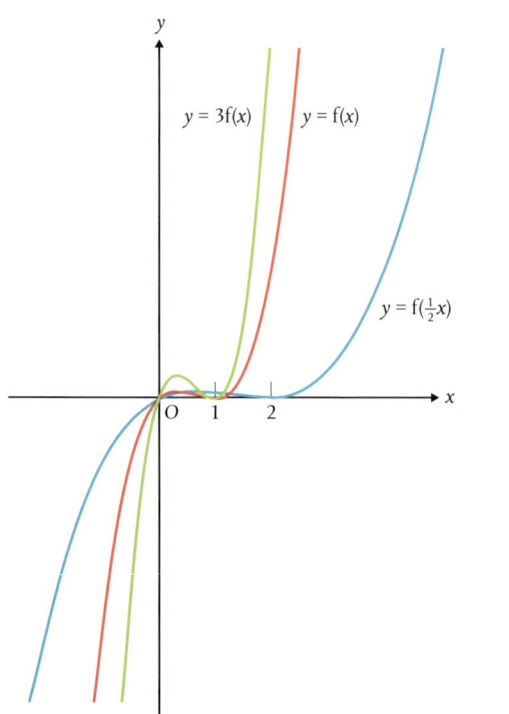

7 a–b

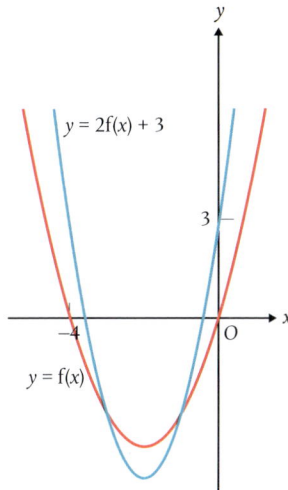

8 a–c

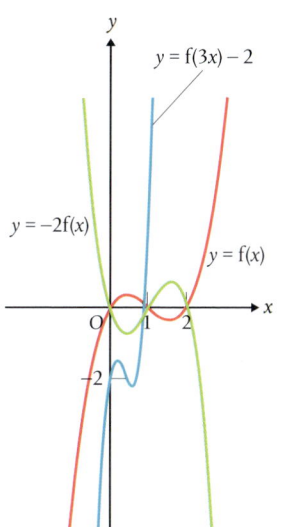

9 a–c

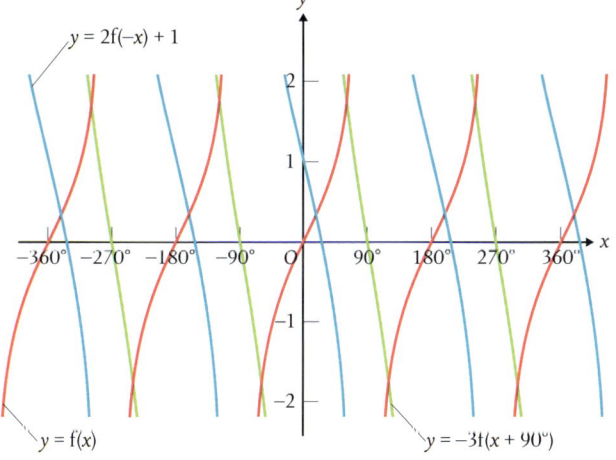

10 a

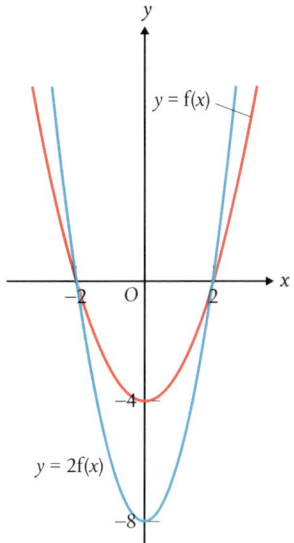

b (0, −4) and (0, −8)

11 a −2, 0 and 3

b −1, 0 and $\frac{3}{2}$

12 a 0 and 6

b 0 and 3

13 a 0 and 1

b 0 and $-\frac{1}{3}$

14 $C_1: y = 2f(x) + 3$

$C_2: y = \frac{1}{2}f(2x + 90°)$

15 a

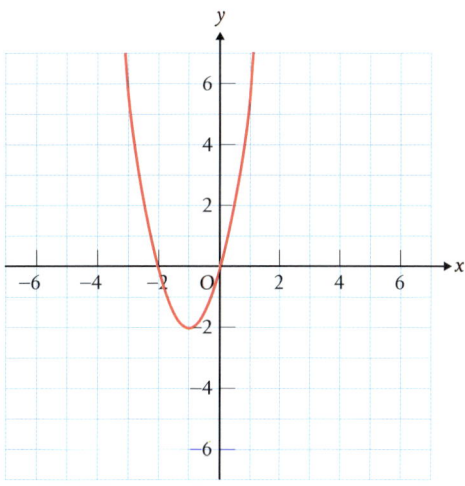

b 24, 8, 0, 0, 8

16 $a = 3$ and $b = -2$

Answers

4.3 Circles

1
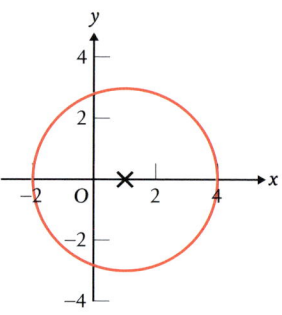

2
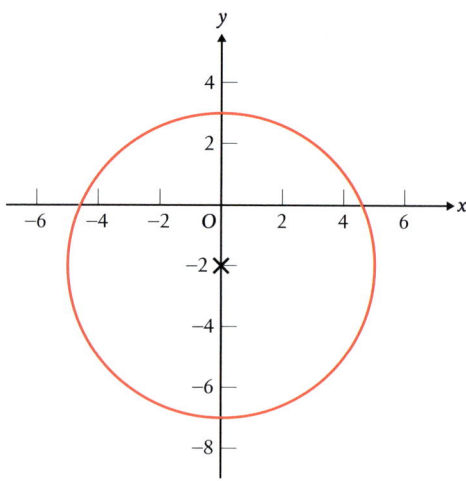

3 $(x-4)^2 + (y+2)^2 = 36$

4
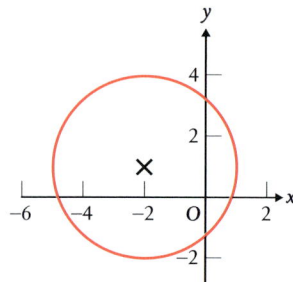

5 a $(2, 1)$
  b $\sqrt{13}$
  c $(x-2)^2 + (y-1)^2 = 13$

6 $(x+1)^2 + (y-7)^2 = 6.25$

7 a Centre $(5, 0)$, radius $\sqrt{30}$
  b Centre $(-3, 0)$, radius $\sqrt{7}$
  c Centre $(2, 0)$, radius 2

8 a Centre $(0, 4)$, radius 4
  b Centre $(0, -1)$, radius $\sqrt{6}$
  c Centre $(0, -2)$, radius $\sqrt{17}$

9 a Centre $(1, -3)$, radius $2\sqrt{5}$
  b Centre $(-3, 2)$, radius $\sqrt{13}$
  c Centre $(-1, 1)$, radius 2

10 Centre $(3, -7)$, radius $\sqrt{11}$

11 a $3^2 + (6-5)^2 = 9 + 1 = 10$, so $(3, 6)$ lies on the circle.
   b $y = \frac{x}{3} + 5$

12 $x - 2y + 5 = 0$

13 $4x + 3y - 52 = 0$

14 $3x + 4y + 37 = 0$

15 $(x-1)^2 + (y-2)^2 = 25$ has centre $(1, 2)$
Radius joining $(1, 2)$ and $(5, 5)$ has gradient $\frac{5-2}{5-1} = \frac{3}{4}$
Tangent at $(5, 5)$ has gradient $-\frac{4}{3}$
$y = -\frac{4}{3}x + c$
$5 = -\frac{4}{3} \times 5 + c$
$c = 5 + \frac{20}{3} = \frac{35}{3}$
Equation of tangent is $y = -\frac{4}{3}x + \frac{35}{3}$, $3y + 4x = 35$

4.4 Exponential and reciprocal graphs

Answers

1 a
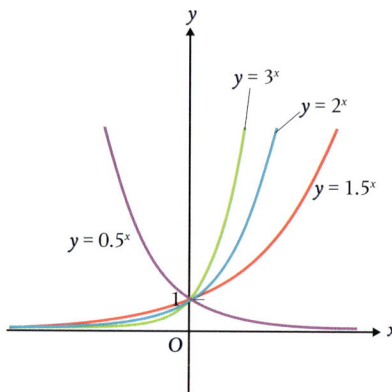

b The graph of $y = 2^x$ is a reflection of the graph of $y = 0.5^x$ in the y-axis.

2 a
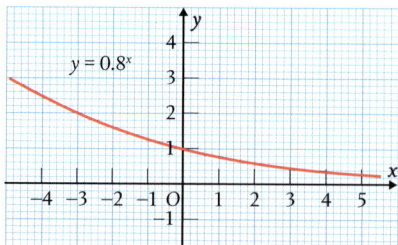

b $x = -3.1$

3 $a = 5$

4 $a = 4$

Answers

5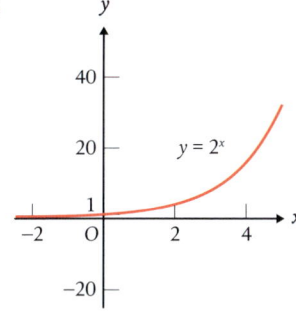

Asymptote: $y = 0$

Intercept: $(0,1)$

6 a–c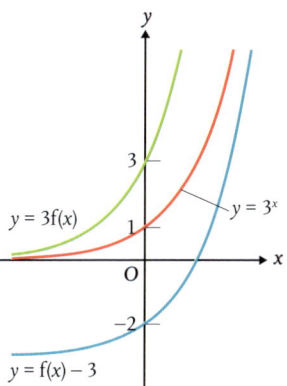

7 As x decreases, y increases

8 a–c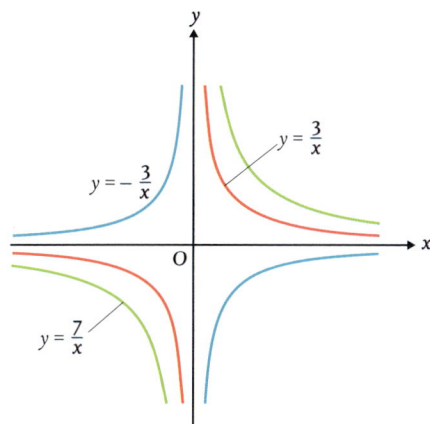

9 a–c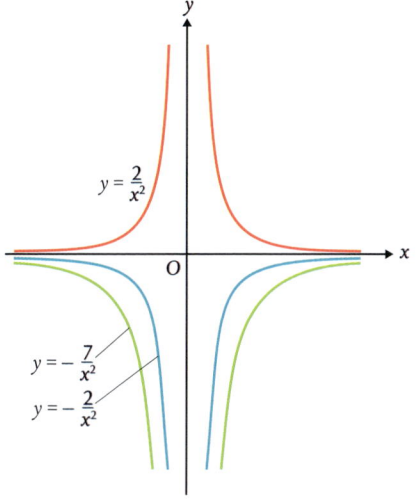

10 a–c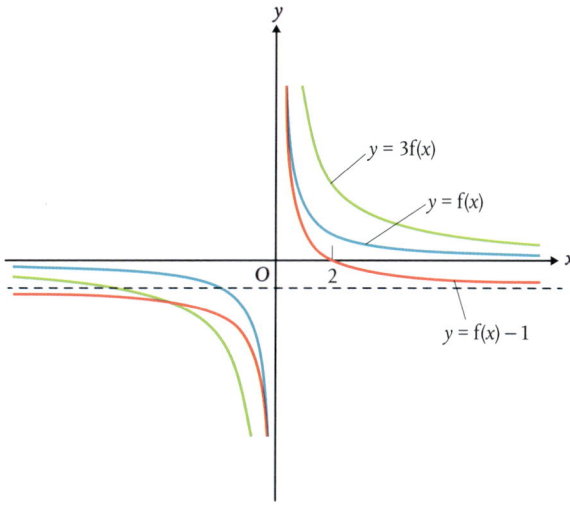

  a Asymptotes: $x = 0, y = 0$

  b Asymptotes: $x = 0, y = -1$

  c Asymptotes: $x = 0, y = 0$

11 a–c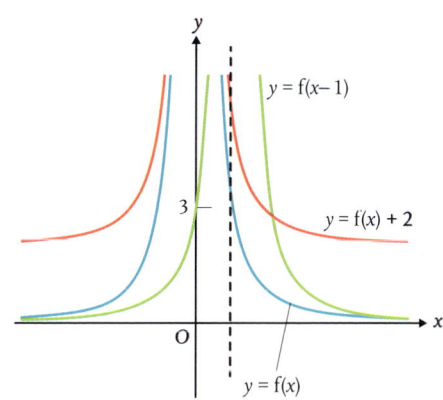

  a Asymptotes: $x = 0, y = 0$

  b Asymptotes: $x = 0, y = 2$

  c Asymptotes: $x = 1, y = 0$

Answers

12 a $x = 0, y = 0$

b $x = 0, y = 1$

c $x = -2, y = 0$

13 a $x = 0, y = -5$

b $x = 0, y = -5$

c $x = -4, y = 0$

14 A i, B iii, C ii

15

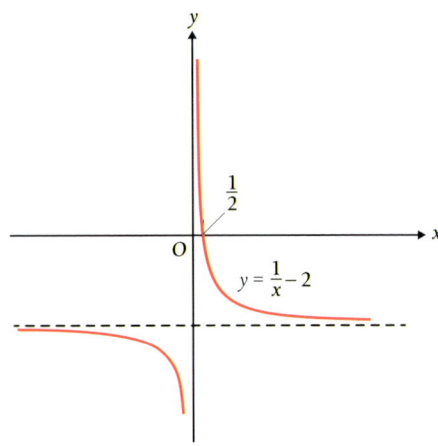

16 $a = 3$ and $b = 0.5$

4.5 Non-linear graphs

1 A: -3, B: 2

2 A: -3, B: 12

3 a

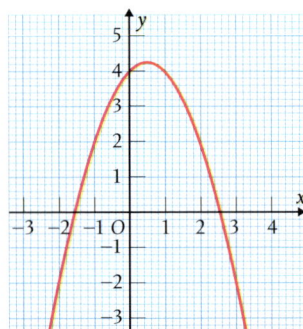

b Approximately -1

c Approximately 3

4 a Approximately 30 m/s

b The speed gradually increases for the first 8 seconds and then gradually slows down.

5 a Approximately 2 m/s^2

b Acceleration

6 37.5 square units

7 24 square units

8 a 12, 12, 10, 6

b 31 square units

9 a 1, 3, 9, 27 and 81

b 80 square units

10 43.6 square units

11 30 square units

12 26 square units

13 80 square units

Chapter 5: Functions

Maths challenge: 5 (number, number of letters in the number)

5.1 Functions

1 a function $f(x) = x + 3$

b not a function

c function $f(x) = \frac{1}{x}$

d function $f(x) = 0$

e function $f(x) = x^4$

f function $f(x) = \sin x$

2 a

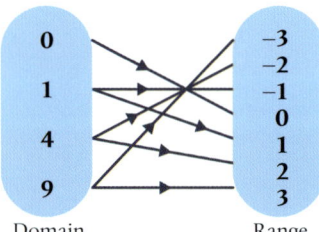

b No, it is a one-to-many relation; some input values map to more than one output value.

3 a 8   b -8 c 60

d 3   e 52   f 70

4 $a = 2, a = 3$

5 $a = -1, a = -5$

6 $a = -2 + \sqrt{22}, a = -2 - \sqrt{22}$

7 a

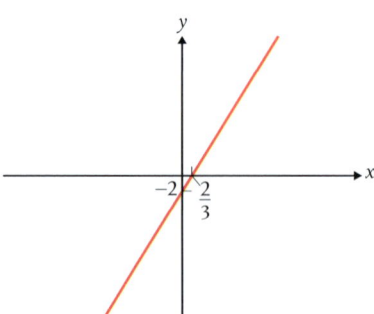

b $\{x \in \mathbb{R} : -4 \leq x \leq 4\}$

c $\{f(x) \in \mathbb{R} : -14 \leq f(x) \leq 10\}$

8 a $\{f(x) \in \mathbb{R} : 0 \leq f(x) \leq 16\}$

b $\{f(x) \in \mathbb{R} : f(x) \geq -6\}$

9 Domain $\{x \in \mathbb{R} : x \neq 0\}$

Range $\{f(x) \in \mathbb{R} : f(x) \neq 0\}$

10 a Domain $\{x \in \mathbb{R} : 0 \leq x < 90° \cup 90° < x < 270° \cup 270° < x \leq 360°\}$

b Range $\{f(x) \in \mathbb{R}\}$

11 a Range $\{y \in \mathbb{R}: -1 \leq y \leq 23\}$

b Range $\{y \in \mathbb{R}: -3 \leq y \leq 7\}$

c $\{y \in \mathbb{R}\}$

12 a Because the domain is integer values of x only.

b The y-values include integers and fractions, so the range is $y \in \mathbb{R}$

13 a Range $\{f(x) \in \mathbb{R} : f(x) \geq -7\}$

b Range $\{f(x) \in \mathbb{R} : f(x) \geq 4\frac{7}{8}\}$

14 $g(x) \geq -7$

15 Range $\{f(x) \in \mathbb{R} : -32 \leq f(x) \leq 34\}$

16 a Range $\{f(x) \in \mathbb{R} : -1 \leq f(x) \leq 1\}$

b Range $\{f(x) \in \mathbb{R} : -1 \leq f(x) \leq 1\}$

c Range $\{f(x) \in \mathbb{R} : -2 \leq f(x) \leq 2\}$

d Range $\{f(x) \in \mathbb{R} : -1 \leq f(x) \leq 1\}$

17 $y = 0$ $0 \leq x < 1$

$y = 2$ $1 \leq x < 2$

$y = 5$ $2 \leq x < 4$

$y = 10$ $4 \leq x < 6$

18 $y = 40$ $0 < x \leq 1.5$

$y = 20x + 10$ $1.5 < x \leq 8$

19 a Domain $\{x \in \mathbb{R} : x \neq 0\}$

b Domain $\{x \in \mathbb{R} : x \neq 4\}$

c Domain $\{x \in \mathbb{R} : x \neq -2, x \neq 3\}$

d Domain $\{x \in \mathbb{R} : x \neq \pm 5\}$

20 $x = 7$

5.2 Composite functions

1 a 9

b 17

c 121

2 a i $\frac{3}{2}$

ii $\frac{1}{9}$

iii 13

b $hg(2) = h(g(2))$

$g(2) = \frac{3}{2-2} = \frac{3}{0}$, which is undefined.

3 a i $fg(x) = 2x + 7$

ii $gf(x) = 2x + 3$

b

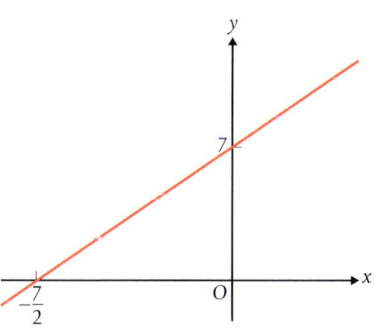

4 a i $gh(x) = x^2 + 4x + 3$

ii $hg(x) = x^2 + 10x + 21$

b

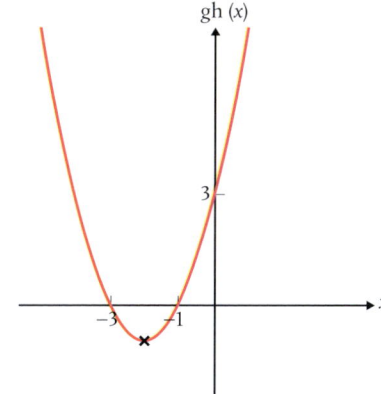

5 a i $pq(x) = (x-1)^3$

ii $qp(x) = x^3 - 1$

b i

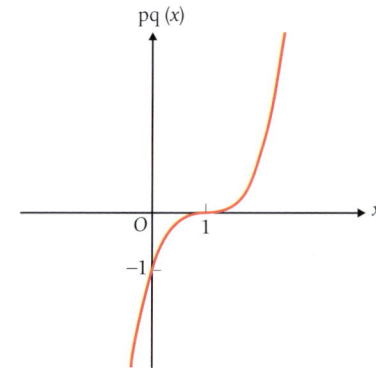

ii

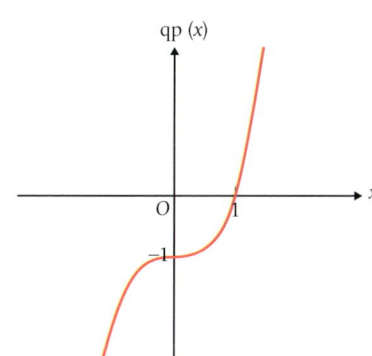

6 Proof shown

7 a i $gh(x) = 3^{2x+1}$ **ii** $hg(x) = 2 \times 3^x + 1$

b i

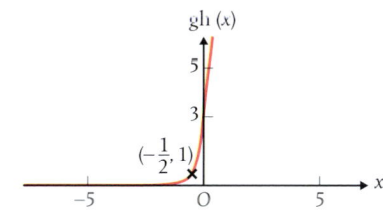

Answers

ii

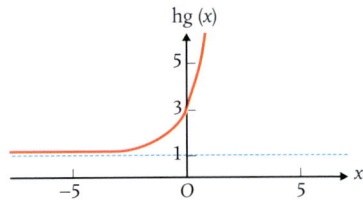

8 **a** fg(x) = cos(x − 30°)

b Range {fg(x) ∈ ℝ : −1 ≤ fg(x) ≤ 1}

9 **a** jh(x) = sin 3x, Range {jh(x) ∈ ℝ : −1 ≤ jh(x) ≤ 1}

b hj(x) = 3 sin x, Range {hj(x) ∈ ℝ : −3 ≤ hj(x) ≤ 3}

10 fg(x) = f(2x) = tan 2x

fg(45°) = tan 90° which is undefined.

11 **a** {x ∈ ℝ : x ≠ −2, x ≠ 1}

b {x ∈ ℝ : x ≠ ±3}

12 **a** p(1) = 1 − 3 + 2 = 0, so using the factor theorem, x − 1 is a factor of x^3 − 3x + 2

b (x − 1)^2(x + 2)

c {x ∈ ℝ : x ≠ 1, x ≠ −2}

5.3 Inverse functions

1 **a** $f^{-1}(x) = \frac{x-4}{5}$

b $g^{-1}(x) = \frac{x+1}{2}$

c $h^{-1}(x) = 2(x-3)$

2 **a** $f^{-1}(x) = \frac{x}{3} + 2$

b $g^{-1}(x) = 2x + 5$

c $h^{-1}(x) = \frac{5x+4}{3}$

3 **a** **i** $f^{-1}(10) = -1$ **ii** $g^{-1}(2) = 9$ **iii** $g^{-1}(-1) = 0$

b $f^{-1}(x) + g^{-1}(x) = \frac{16x}{5}$

4 **a** $f^{-1}(x) = -x$

b $g^{-1}(x) = 1 - x$

c $h^{-1}(x) = x$

All three functions are self-inverse.

5 **a** **i** $\frac{1}{2}$ **ii** 2 **iii** $\frac{1}{n}$ **iv** n

b Find the reciprocal

c $h^{-1}(x) = \frac{1}{x} = h(x)$

d {x ∈ ℝ : x ≠ 0}

6 **a** $f^{-1}(x) = \frac{3x}{2} - 1$

b $g^{-1}(x) = \frac{4x}{3} - 1$

c $h^{-1}(x) = \frac{1}{3}\left(\frac{5x}{2} - 1\right)$

7 2

8 a–b

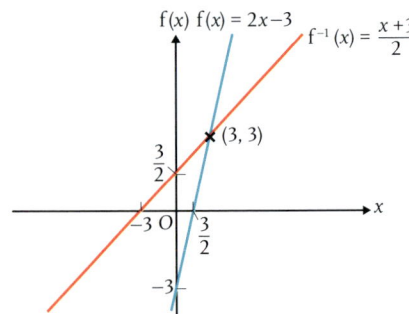

c {x ∈ ℝ}

9

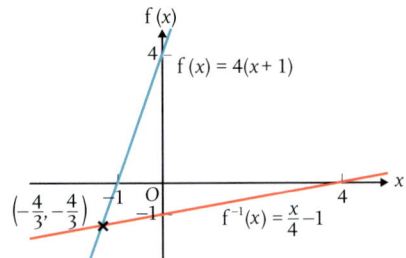

10 a–b

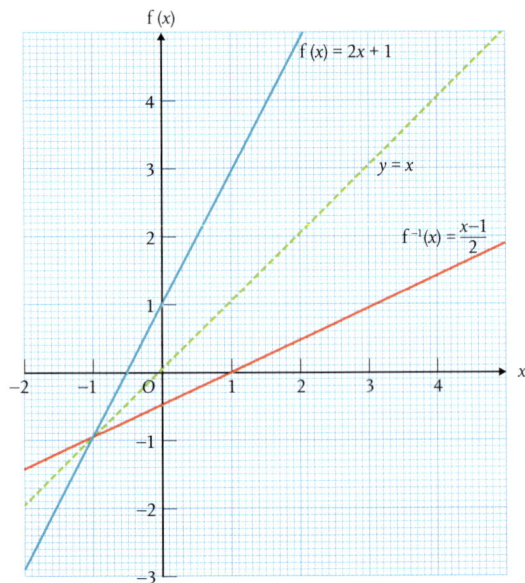

c Reflection in the line $y = x$

d The graph of f(x) is a reflection of $f^{-1}(x)$ in the line $y = x$

The intersection of the graphs of f(x) and $f^{-1}(x)$ lies on the line $y = x$

11 **a** $\sqrt{4} = \pm 2$

b

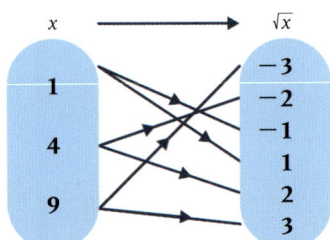

Answers

c $x \to \sqrt{x}$ is a one-to-many relation on the range $\{\sqrt{x} \in \mathbb{R}\}$, so it is not a function.

d
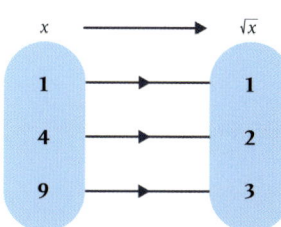

e $x \to \sqrt{x}$ is a one-to-one relation on the range $\{\sqrt{x} \in \mathbb{R} : \sqrt{x} \geq 0\}$ so it is a function.

12 a $f^{-1}(x) = \sqrt{x}$, domain $\{x \in \mathbb{R} : x \geq 0\}$, range $\{f^{-1}(x) \in \mathbb{R} : f^{-1}(x) \geq 0\}$

b $g^{-1}(x) = \sqrt[3]{x}$, domain $\{x \in \mathbb{R}\}$

c $h^{-1}(x) = \sqrt[4]{x}$, domain $\{x \in \mathbb{R} : x \geq 0\}$, range $\{h^{-1}(x) \in \mathbb{R} : h^{-1}(x) \geq 0\}$

13
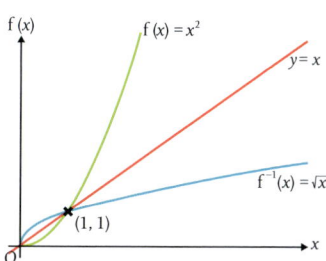

14 a $f^{-1}(x) = x^2$

b $g^{-1}(x) = (x-3)^2$

c $h^{-1}(x) = (x-5)^2 - 1$

15 a $fg(x) = x$

b f and g are inverse functions. $f^{-1}(x) = g(x)$ and $f(x) = g^{-1}(x)$

16
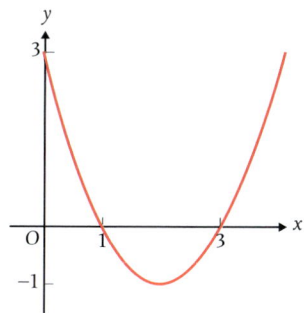

5.4 Transforming functions

1 a $y = x^2 - 2$ **b** $y = (x-2)^2$ **c** $y = 4x^2$ **d** $y = -2x^2$

2 a $y = 3x - 1$ **b** $y = 3x + 4$ **c** $y = -6x - 2$ **d** $y = 6x - 4$

3 a $y = x^2 + 2$ **b** $y = x^2 + 2x + 6$ **c** $y = 9x^2 + 5$ **d** $y = -2x^2 - 10$

4 a $y = x^2 - 2x + 2$ **b** $y = x^2 - 6x + 8$ **c** $y = 4x^2 + 4x$ **d** $y = 4x^2 - 8x$

5 a $y = x^3 - x + 1$ **b** $y = x^3 - 3x^2 + 2x$

c $y = -8x^3 + 2x$ **d** $y = -2x^3 + 2x$

6 a $y = 4x$ **b** $y = 6x + 15$ **c** $y = -4x + 7$

7 a $y = 9x^2 + 1$ **b** $y = 2x^2 - 4x + 2$ **c** $y = 4x^2 + 4x + 1$

8 a $y = 2x^2 + 2x$ **b** $y = 9x^2 - 9x + 2$ **c** $y = 4x^2 - 2x + 1$

9 a $y = 4^{2x-3}$ **b** $y = 4^{2x} + 3$ **c** $y = 2 \times 4^{x+3}$

10 a $y = \dfrac{1}{2x^2} - 1$ **b** $y = \dfrac{4}{(x-1)^2}$ **c** $y = \dfrac{2}{(-2x+1)^2}$

11 a $y = x^3$ **b** $y = x^3 - 4$

12 a $y = x^2$ **b** $y = (x-3)^2$

13 a $y = \cos x$ **b** $y = -3\cos x - 1$

14 A iii, B vi, C i, D v, E ii, F iv

15 $y = \dfrac{1}{(x-2)^2} - 1$

Chapter 6: Equations and inequalities

Maths challenge: 243 (the digits are cubed and added together)

6.1 Solve equations

1 a $x = 4$
b $x = 6$
c $x = 5$
d $x = -3$

2 a $x = 2$
b $x = -\dfrac{37}{13}$

3 a $x = \dfrac{9}{7}$
b $x = 2\dfrac{3}{5}$
c $x = \dfrac{13}{16}$
d $x = -1$

4 a $R_2 = 30$
b $R = \dfrac{15}{23}$
c $R_1 = 16, R_2 = 48$

5 a $f = \dfrac{5v}{4}$
b $u = 15, v = 7.5$

6 a $x = 8$
b $x = -3$
c $x = 5$
d $x = 3$

7 $x = 14.5°, x = 165.5°$

8 $x = 30°, x = 150°$

9 $x = -60°, x = 60°$

10 $\theta = 74.1°, \theta = 254.1°$

11 $\pm 45°$

12 a full correct sketch with -1 and 1 marked on the y-axis

13 a equations i and vi

b ii $x = 116.6°, 63.4°$

iii $x = 0°$

iv $x = -150°, x = -30°$

v $x = -30°, x = 30°$

14 a $y = 2\cos x + 5$. When $x = 30°, y = 5 + \sqrt{3}$.

b When $y = 0$, $2\cos x + 5 = 0$, $2\cos x = -5$, $\cos x = -\dfrac{5}{2}$, which has no solutions because $\cos x$ must be between -1 and 1

Answers

15 a $a = 10$
b $a = \frac{1}{2}$
16 $a = 139°$ or $221°$
17 $a = 21$

6.2 Solve quadratic equations

1 a $x = \frac{-5 + \sqrt{57}}{4}$ and $x = \frac{-5 - \sqrt{57}}{4}$
b $x = \frac{2 - \sqrt{14}}{5}$ and $x = \frac{2 + \sqrt{14}}{5}$
c $x = \frac{-3 - \sqrt{17}}{4}$ and $x = \frac{-3 + \sqrt{17}}{4}$

2 a $x = -3$ and $x = \frac{2}{3}$
b $x = \frac{-1 - \sqrt{2}}{2}$ and $x = \frac{-1 + \sqrt{2}}{2}$
c $x = \frac{3 - \sqrt{105}}{12}$ and $x = \frac{3 + \sqrt{105}}{12}$

3 57 and 59

4 729 and 784

5 6.2 cm and 10.2 cm

6 $\frac{1}{2}bh = \frac{1}{2} \times (x + 7)2x = x^2 + 7x = 35$
$x^2 + 7x - 35 = 0$
$x = \frac{-7 \pm \sqrt{49 - 4 \times -35}}{2}$
$x = \frac{-7 \pm \sqrt{189}}{2}$
$x = \frac{-7 + 3\sqrt{21}}{2}$
(ignoring the negative solution, because $2x$ is a length, so x must be positive)
Therefore, the height of the triangle is $-7 + 3\sqrt{21}$ metres

7 a The height of the cable is 3 m at the centre of the bridge.
b 80 metres

8 a 120 m
b Max height (where $x = 60$) = 9 m

9 a i $fg(x) = 2x^2 - 1$ **ii** $gf(x) = 4x^2 - 12x + 10$
b $x = \frac{6 \pm \sqrt{14}}{2}$

10 a $x = 2$ and $x = -\frac{9}{4}$
b $x = -1$ and $x = \frac{14}{3}$
c $x = -3$ and $x = 2$
d $x = \frac{13 + \sqrt{213}}{2}$ and $x = \frac{13 - \sqrt{213}}{2}$

11 a $\frac{5}{x - 2} + \frac{3x + 1}{2x} = 2$
$\frac{10x + (3x + 1)(x - 2)}{(x - 2)2x} = 2$
$10x + 3x^2 + x - 6x - 2 = 4x(x - 2)$
$3x^2 + 5x - 2 = 4x^2 - 8x$
$x^2 - 13x + 2 = 0$
b $x = \frac{13 + \sqrt{161}}{2}$ and $x = \frac{13 - \sqrt{161}}{2}$

12 a $\frac{5}{2x + 1} + \frac{3}{x - 2} = 1$
$\frac{5(x - 2) + 3(2x + 1)}{(2x + 1)(x - 2)} = 1$
$5x - 10 + 6x + 3 = (2x + 1)(x - 2)$
$11x - 7 = 2x^2 - 3x - 2$
$2x^2 - 14x + 5 = 0$
b Solutions are $x = 0.378$ and $x = 6.62$, which are between 0 and 7.

13 a $x^2 - 5x - 2 = 0$
b $\frac{5 \pm \sqrt{33}}{2}$

14 a $u^2 - 7u + 12 = 0$
b $u = 3$ and $u = 4$
c $x = \pm\sqrt{3}$ and $u = \pm 2$

15 $x = 9$ or $x = 4$

16 $x = -90°$ and $x = 30°$ and $x = 150°$

17 a $x = -\frac{1}{2}$ or $x = 1$ **b** $x = 0°, 120°$ or $240°$

6.3 Solve simultaneous equations

1 a $x = 3, y = 2$
b $x = -1, y = 4$
c $x = 7.14, y = 5.86$ (2 d.p.)

2 a $x + y = 2.3$
b time = $\frac{\text{distance}}{\text{speed}}$
  i The time taken to swim x km front crawl at a speed of 3.2 km/h = $\frac{x}{3.2}$ hours
  ii The time taken to swim y km breaststroke at a speed of 2 km/h = $\frac{y}{2}$ hours
c $\frac{x}{3.2} + \frac{y}{2} = 1$
d 1.5 km

3 $\frac{5}{9}$

4 $a = 2, b = -5$

5 a $x = -\frac{1}{2}, y = 3\frac{1}{4}$ and $x = 0, y = 3$
b $x = 0, y = 1$ and $x = 2, y = 5$
c $x = 1, y = 5$ and $x = -\frac{1}{2}, y = 3\frac{1}{2}$

6 a area of triangle = $\frac{1}{2}bh = \frac{1}{2}x(2x + 3) = 6$
$x(2x + 3) = 12$
$2x^2 + 3x = 12$
x is 2 cm less than y, $y = x + 2$
b $y = 3.8$ cm (to 1 d.p.)

7 (3, 3)

8 a $x = 1, y = 3$
b $x = 1, y = 4$ and $x = \frac{5}{4}, y = 5$
c $x = -4, y = 13$ and $x = -\frac{5}{3}, y = 6$

9 $(-1 - \sqrt{2}, 5 - \sqrt{2})$ and $(-1 + \sqrt{2}, 5 + \sqrt{2})$

10 a $x = -4, y = 3$ and $x = 4.8, y = -1.4$

b

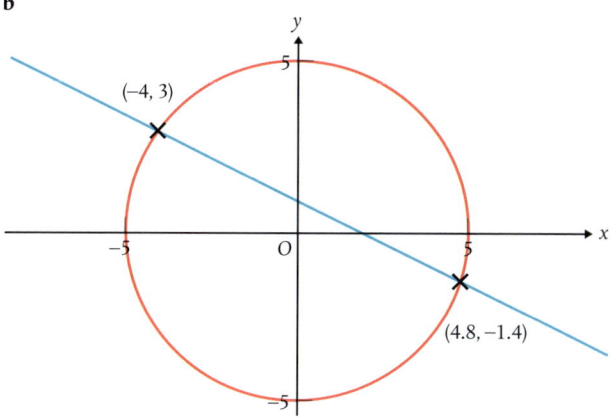

11 $x^2 + y^2 = 16$ and $y = 2, (-2\sqrt{3}, 2)$ and $(2\sqrt{3}, 2)$

12 $(-3, -2)$ and $(2, 3)$

13 $\left(\dfrac{10 - 2\sqrt{5}}{5}, \dfrac{5 + 4\sqrt{5}}{5}\right)$ and $\left(\dfrac{10 + 2\sqrt{5}}{5}, \dfrac{5 - 4\sqrt{5}}{5}\right)$

14 $x^2 + y^2 = 10$

$y = 3x + 10$

$x^2 + (3x + 10)^2 = 10$

$x^2 + 9x^2 + 60x + 100 = 10$

$10x^2 + 60x + 90 = 0$

$x^2 + 6x + 9 = 0$

$(x + 3)(x + 3) = 0$

$x = -3$

The only solution is $x = -3, y = 1$, so they intersect at the point $(-3, 1)$

b The line $y = 3x + 10$ is a tangent to the circle

15

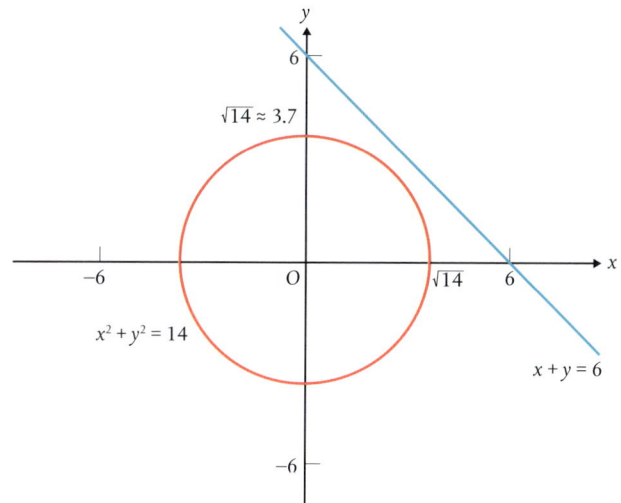

16 $(0, 1)$ and $(-2, 3)$

17 $(5.73, 3.45)$ $(3.07, -1.85)$

6.4 Solve inequalities

1 a $x < -4$

b $x > \dfrac{4}{5}$

c $x \leq 6$

2 a $x \geq 30, y \geq 2x$

b Total number of hectares for wheat and maize is greater than 0 and up to a maximum 140 hectares

c

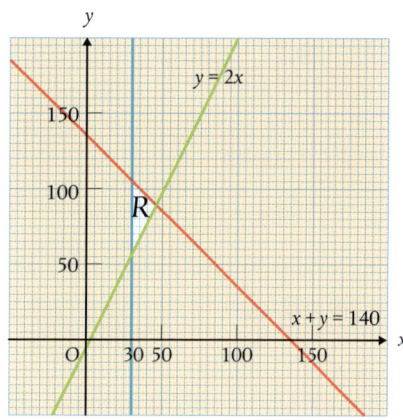

3 a $x = 0, y = 1$

b $x = 1.5, y = 3.5$

4

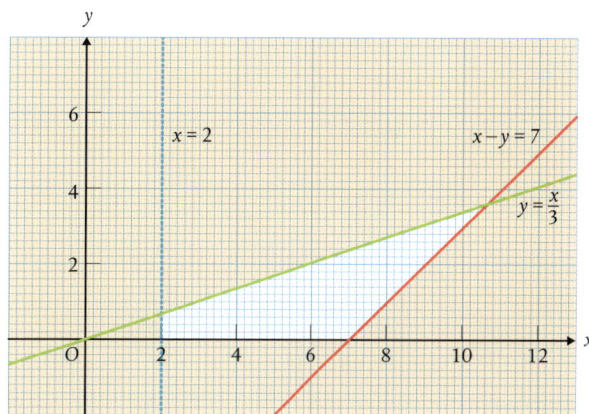

$x = 10.5, y = 3.5$

5 a $x \leq 12$ and $y \leq 15$

b the total number of cars hired is greater than 5 and less than 21

Answers

c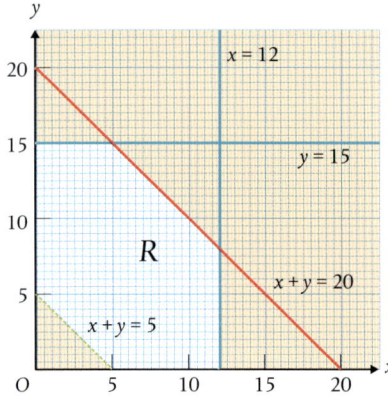

d You cannot hire a negative number of cars in a day.

e $75x + 105y$ (= Charge)

f 1950

6 a $T = 32x + 20y$

b $x = 200, y = 200$ (200 adult tickets, 200 child tickets)

7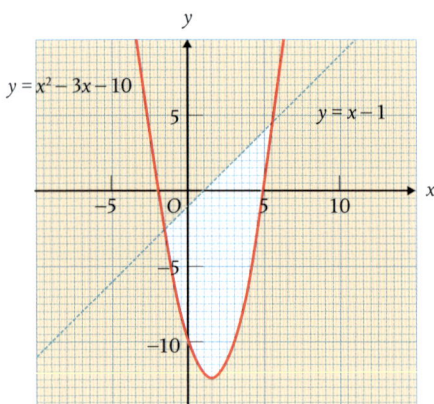

8 a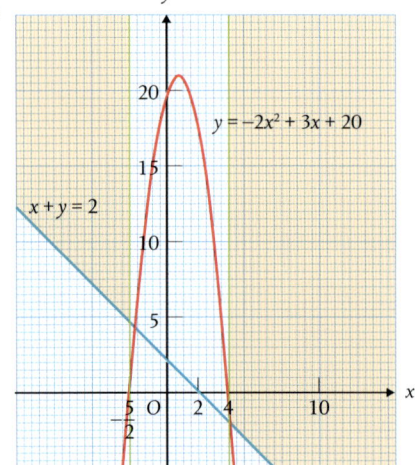

b $\left\{(x, y): -\dfrac{5}{2} \leq x \leq 4 \text{ and } x + y \leq 2\right\}$

9 a $\{x: -1 < x < -0.4\}$

b $\{x: -1 \leq x \leq -0.4\}$

c $\{x: x < -1\} \cup \{x: x > -0.4\}$

d $\{x: x \leq -1\} \cup \{x: x \geq -0.4\}$

10 a and **c**

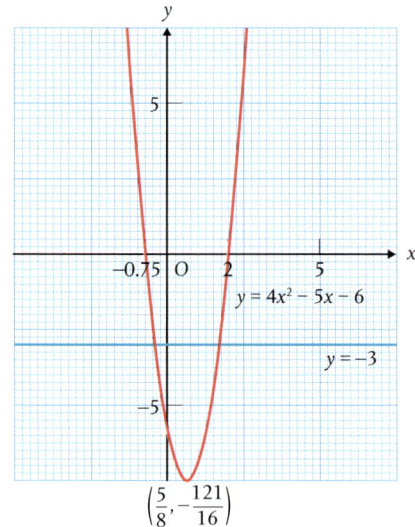

b $\{x: -0.75 \leq x \leq 2\}$

c $\{x: -0.5 \leq x \leq 1.75\}$

11 a $\{x: x < -2\} \cup \{x: x > 1\}$

b $\left\{x: x \leq \dfrac{1}{3}\right\} \cup \{x: x \geq 4\}$

c $\{x: 1 \leq x \leq 4\}$

12 a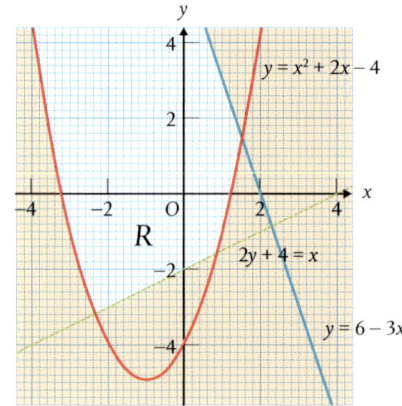

b 1.53

Answers

Chapter 7: Pythagoras and trigonometry

Maths challenge: $w = 13, x = 5, y = 18, z = 8$

7.1 Pythagoras and trigonometry
1. a 6.19 cm
   b 8.69 cm
   c 5.64 cm
2. a 7.27 cm
   b 5.82 cm
   c 5.41 cm
3. a 42.5°
   b 19.1°
   c 17.9°
4. 9.45 cm
5. $x\sqrt{6}$
6. $8x$ cm
7. 35.7°
8. 64.1°
9. 70.9°
10. 5.59 square units
11. Correct triangle drawn and then correct ratio shown and substituted

7.2 Sine and cosine rules, and area of a triangle
1. a 1.55 cm
   b 21.0 cm
   c 8.76 cm
2. a 36.2°
   b 27.4°
   c 43.8°
3. a 21.0 cm²
   b 12.4 cm²
   c 21.5 cm²
4. 108°
5. 41.9°
6. $2(\pi - 2\sqrt{2})$
7. 104
8. 28.1 cm²
9. 1.41 cm²
10. 614 cm³
11. 1.04 cm³

Chapter 8: Probability

Maths challenge: $10 \times \sqrt{2} - 10 = 4.14$

8.1 The language of probability
1. a

	Art	Music	Drama	Total
Year 7	8	13	9	30
Year 8	0	2	9	11
Total	8	15	18	41

   b $\frac{13}{41}$
   c $\frac{9}{18} = \frac{1}{2}$
2. a $\frac{1}{3}, \frac{2}{3}, \frac{1}{3}, \frac{2}{3}, \frac{1}{3}, \frac{2}{3}$
   b $\frac{1}{9}$
   c $\frac{4}{9}$
3. a 0.3, 0.5, 0.2, 0.3, 0.5, 0.2, 0.3, 0.5, 0.2, 0.3, 0.5, 0.2
   b 0.38
4. Berat should have added not multiplied.
5. a

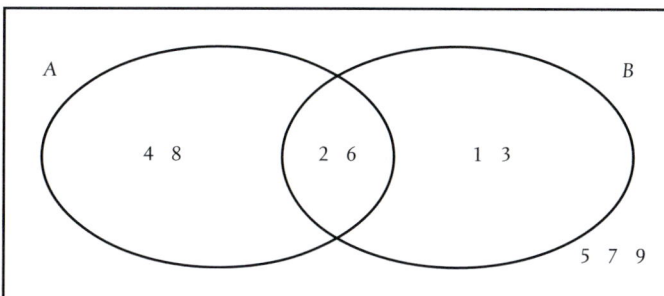

   b $\frac{2}{9}$
   c $\frac{6}{9} = \frac{2}{3}$
   d $\frac{5}{9}$
6. a 29
   b i $\frac{8}{120} = \frac{1}{15}$
      ii $\frac{26}{38} = \frac{13}{19}$
7. a

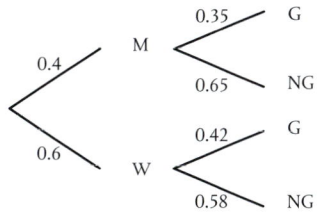

   b 0.608
8. 0.77
9. 0.08

Answers

10 a

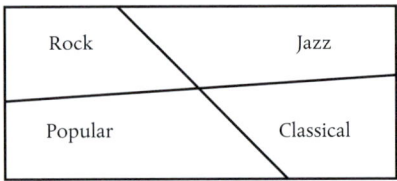

b 0.7

11 $P(A \cup B)$
$= \frac{z + y - x}{n}$
$= \frac{z}{n} + \frac{y}{n} - \frac{x}{n}$
$= P(A) + P(B) - P(A \cap B)$

8.2 Probability problems

1 a

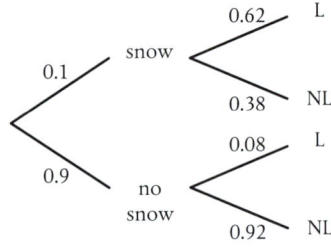

b 0.072

2 a

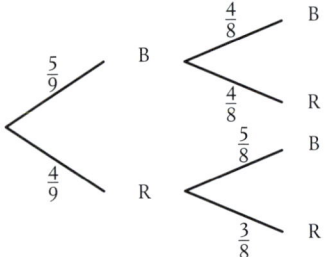

b $\frac{32}{72} = \frac{4}{9}$

3 a

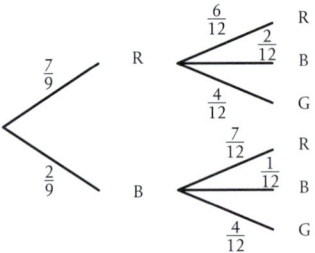

b $\frac{64}{108} = \frac{16}{27}$

4 $\frac{12x - x^2}{66}$

5 a Shown   b $\frac{14}{39}$

6 0.35

7 139

8 15

Chapter 9: Proof

Maths challenge: 18 (the numbers are all 18 in base 2, base 3, base 4 etc)

9.1 Proof by deduction

1 $(n + 1)^2 - n^2 = n^2 + 2n + 1 - n^2$
$= 2n + 1$
$= n + (n + 1)$

2 $(n \times (n + 1)) + (n + 1) = n^2 + n + n + 1$
$= n^2 + 2n + 1$
$= (n + 1)^2$

3 $(2n + 1)^2 - (2n + 1) = (4n^2 + 4n + 1) - (2n + 1)$
$= 4n^2 + 2n$
$= 2(2n^2 + n)$

4 $(2n - 1)^2 + (2n + 1)^2 + (2n + 3)^2$
$= (4n^2 - 4n + 1) + (4n^2 + 4n + 1) + (4n^2 + 12n + 9)$
$= 12n^2 + 12n + 11$
$= 12(n^2 + n) + 11$

5 $(4n - 1)^2 - (4n - 5)^2$
$= (16n^2 - 8n + 1) - (16n^2 - 40n + 25)$
$= 32n - 24$
$= 8(4n - 3)$

6 $\frac{x}{1 + \sqrt{12}} = \frac{x}{1 + 2\sqrt{3}}$
$= \frac{x}{1 + 2\sqrt{3}} \times \frac{1 - 2\sqrt{3}}{1 - 2\sqrt{3}}$
$= \frac{x - 2x\sqrt{3}}{1 + 2\sqrt{3} - 2\sqrt{3} - 12}$
$= \frac{x - 2x\sqrt{3}}{-11}$
$= \frac{2x\sqrt{3} - x}{11}$

7 $(2n + 1) \times (2m + 1) = 4nm + 2n + 2m + 1$
$= 2(2nm + n + m) + 1$

8 $3x^2 + 24x + 56 = 3(x^2 + 8x) + 56$
$= 3((x + 4)^2 - 16) + 56$
$= 3(x + 4)^2 + 8$

Minimum at $(-4, 8)$

9 $2x^2 - 6x + \frac{11}{2} = 2(x^2 - 3x) + \frac{11}{2}$
$= 2\left(\left(x - \frac{3}{2}\right)^2 - \frac{9}{4}\right) + \frac{11}{2}$
$= 2\left(x - \frac{3}{2}\right)^2 + 1$

Minimum at $\left(\frac{3}{2}, 1\right)$

10 $10x - 5x^2 - 12 = -5(x^2 - 2x) - 12$
$= -5((x - 1)^2 - 1) - 12$
$= -5(x - 1)^2 - 7$

Maximum at $(1, -7)$

11 $\left(x + \frac{1}{x}\right)^3 = x^3\left(\frac{1}{x}\right)^0 + 3x^2\left(\frac{1}{x}\right)^1 + 3x^1\left(\frac{1}{x}\right)^2 + x^0\left(\frac{1}{x}\right)^3$
$= x^3 + 3x + \frac{3}{x} + \frac{1}{x^3}$

Answers

12 $\dfrac{(n+2)^2 - (n+1)^2}{2n^2 + 3n} = \dfrac{(n^2 + 4n + 4) - (n^2 + 2n + 1)}{2n^2 + 3n}$

$= \dfrac{2n + 3}{n(2n + 3)}$

$= \dfrac{1}{n}$

13 $\sin B = \dfrac{h}{c}$ and $\sin C = \dfrac{h}{b}$

$h = c \sin B$ and $h = b \sin C$

$c \sin B = b \sin C$

$\dfrac{c}{\sin C} = \dfrac{b}{\sin B}$

14 **a** $b^2 = x^2 + h^2$

   b $a^2 = (c - x)^2 + h^2$

   c $h^2 = b^2 - x^2$ and $h^2 = a^2 - (c - x)^2$

   $b^2 - x^2 = a^2 - (c - x)^2$

   $b^2 = a^2 - (c - x)^2 + x^2$

   d $\cos A = \dfrac{x}{b}$ so $x = b \cos A$

   e $b^2 = a^2 - (c^2 - 2cx + x^2) + x^2$

   $b^2 = a^2 - c^2 + 2cx$

   $b^2 = a^2 - c^2 + 2cb \cos A$

   $a^2 = b^2 + c^2 - 2bc \cos A$

15 Proof

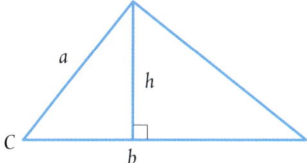

9.2 Proof by exhaustion and disproof by counter example

1 $n = 3$ $3^2 + 2 = 11$ not divisible by 4
   $n = 4$ $4^2 + 2 = 18$ not divisible by 4
   $n = 5$ $5^2 + 2 = 27$ not divisible by 4
   $n = 6$ $6^2 + 2 = 38$ not divisible by 4
   $n = 7$ $7^2 + 2 = 51$ not divisible by 4

   Thus, $n^2 + 2$ is not a multiple of 4 when $3 \leq n \leq 7$.

2 Possible values of n are 5, 7, 11, 13
   $n = 5$ $(5 - 1)(5 + 1) = 24 = 12 \times 2$
   $n = 7$ $(7 - 1)(7 + 1) = 48 = 12 \times 4$
   $n = 11$ $(11 - 1)(11 + 1) = 120 = 12 \times 10$
   $n = 13$ $(13 - 1)(13 + 1) = 168 = 12 \times 14$

   So, $(n - 1)(n + 1)$ is a multiple of 12 when n is a prime number and $3 < n < 14$.

3 Possible values of n are 3, 5, 7, 9, 11, 13, 15, 17, 19
   $n = 3$ prime
   $n = 5$ prime
   $n = 7$ prime
   $n = 9$ $= 3 \times 3$
   $n = 11$ prime
   $n = 13$ prime
   $n = 15$ $= 3 \times 5$
   $n = 17$ prime
   $n = 19$ prime

   So, n is either prime or a product of exactly 2 primes when n is an odd integer such that $2 < n < 20$.

4 If n is not a multiple of 3, it must be either one more than a multiple of 3 or two more than a multiple of 3, e.g.

   $3k + 1$ or $3k + 2$

   $n^2 - 1 = (3k + 1)^2 - 1$
   $= 9k^2 + 6k + 1 - 1$
   $= 3(3k^2 + 2k)$

   $n^2 - 1 = (3k + 2)^2 - 1$
   $= 9k^2 + 12k + 4 - 1$
   $= 3(3k^2 + 4k + 1)$

   So, $n^2 - 1$ is a multiple of 3 if n is not a multiple of 3.

5 If n is 2 more than a multiple of 5, then it must be in the form $5n + 2$ and then m must be $5n + 3$.

   $nm = (5n + 2)(5n + 3)$
   $= 25n^2 + 25n + 6$
   $= 5(5n^2 + 5n + 1) + 1$

   Which is 1 more than a multiple of 5.

6 Possible values of n are 3, 4, 5, 6, 7
   $n = 3$ $3^3 = 27$ $= 9 \times 3$
   $n = 4$ $4^3 = 64$ $= 9 \times 6 + 1$
   $n = 5$ $5^3 = 125$ $= 9 \times 14 - 1$
   $n = 6$ $6^3 = 216$ $= 9 \times 24$
   $n = 7$ $7^3 = 343$ $= 9 \times 38 + 1$

   So, n^3 is either a multiple of 9, or one more or one less than a multiple of 9 when $3 \leq n \leq 7$

7 If n is 1 more than a multiple of 4 it is in the form $4n + 1$, whereas if it is 1 less than a multiple of 4, it is in the form $4n - 1$

   $(4n + 1)^2 = 16n^2 + 8n + 1$
   $= 4(4n^2 + 2n) + 1$

   Which is 1 more than a multiple of 4.

   $(4n - 1)^2 = 16n^2 - 8n + 1$
   $= 4(4n^2 - 2n) + 1$

   Which is 1 more than a multiple of 4.

8 If n is odd it is in the form $2n + 1$.
   Using Pascal's triangle

   $(2n + 1)^4 = 1(2n)^4(1)^0 + 4(2n)^3(1)^1 + 6(2n)^2(1)^2 + 4(2n)^1(1)^3 + 1(2n)^0(1)^4$
   $= 16n^4 + 32n^3 + 24n^2 + 8n + 1$
   $= 8(2n^4 + 4n^3 + 3n^2 + n) + 1$

   Which is 1 more than a multiple of 8.

9 Need to test for each possible unit value.

   if n ends in a 1, then n^2 ends in a 1
   if n ends in a 2, then n^2 ends in a 4
   if n ends in a 3, then n^2 ends in a 9
   if n ends in a 4, then n^2 ends in a 6
   if n ends in a 5, then n^2 ends in a 5
   if n ends in a 6, then n^2 ends in a 6
   if n ends in a 7, then n^2 ends in a 9
   if n ends in a 8, then n^2 ends in a 4
   if n ends in a 9, then n^2 ends in a 1
   if n ends in a 0, then n^2 ends in a 0

   So, no square number ends in an 8.

199

Answers

10 a e.g. $3^2 = 9$ is not even.

b e.g. $5 + -2 = 3$ is not greater than 5.

c e.g. $2 \times 8 = 16$

1 is not even.

d e.g. $x = 7$

$7^2 + 7 - 7 = 49$; 49 is not prime.

e e.g. $a = 1, b = -3$

so, $a > b$

$a^2 = 1^2 = 1, b^2 = (-3)^2 = 9$

so, a^2 is not greater than b^2.

f e.g. $n = 7$ is prime.

$n + 2 = 7 + 2 = 9$ is not prime.

11 $1^2 + 1 + 11 = 13$

$2^2 + 2 + 11 = 17$

$3^2 + 3 + 11 = 23$

$4^2 + 4 + 11 = 31$

$5^2 + 5 + 11 = 41$

12 e.g. $3^2 - 3 + 3 = 9$

9.3 Geometric proof

1 length $AB = \sqrt{(4-1)^2 + (2--1)^2}$

$= 3\sqrt{2}$

length $BC = \sqrt{(7-4)^2 + (-1-2)^2}$

$= 3\sqrt{2}$

$AB = BC$ so triangle is isosceles.

gradient of $AB = \frac{2--1}{4-1} = 1$

gradient of $BC = \frac{-1-2}{7-4} = -1$

gradient of $AB \times$ Gradient of $BC = 1 \times -1 = -1$

So AB and BC are perpendicular and triangle is right angled.

2 length of $AB = \sqrt{(1--1)^2 + (3-2)^2}$

$= \sqrt{5}$

length of $BC = \sqrt{(3-1)^2 + (2-3)^2}$

$= \sqrt{5}$

length of $CD = \sqrt{(1-3)^2 + (-2-2)^2}$

$= 2\sqrt{5}$

length of $DA = \sqrt{(-1-1)^2 + (2--2)^2}$

$= 2\sqrt{5}$

So, $AB = BC$ and $CD = DA$.

Therefore, $ABCD$ is a kite.

3 gradient of $AB = \frac{3-1}{-1--2}$

$= 2$

gradient of $DC = \frac{1-3}{3-4}$

$= 2$

gradient of $BC = \frac{3-3}{4--1}$

$= 0$

gradient of $AD = \frac{1-1}{3--2}$

$= 0$

Therefore, AB and DC are parallel and BC and AD are parallel.

So, $ABCD$ is a parallelogram.

4 $AB = DC$ (opposite sides of a parallelogram are equal)

angle BAC = angle ACD (alternate angles are equal)

angle ABD = angle BDC (alternate angles are equal)

So, using ASA (Angle-Side-Angle), ABE and DCE are congruent.

5 $AB = CD$ (given information)

angle ABC = angle BCD (given information)

BC is a common side in triangles ABC and BCD

So, using SAS (Side-Angle-Side) triangles ABC and BCD are congruent.

Therefore, $AC = BD$

6 $x^2 + 2x + y^2 - 4y - 20 \Rightarrow (x+1)^2 + (y-2)^2 = 25$

$(x+1)^2 + \left(\left(\frac{4}{3}x - 5\right) - 2\right)^2 = 25$

$\frac{25}{9}x^2 - \frac{50}{3}x + 25 = 0$

$x^2 - 6x + 9 = 0$

$(x-3)^2 = 0$

$x = 3$

$y = \frac{4}{3} \times 3 - 5$

$y = -1$

Point of intersection is $(3, -1)$

Gradient of radius to point $(3, -1) = \frac{-1-2}{3--1} = -\frac{3}{4}$

$-\frac{3}{4} \times \frac{4}{3} = -1$

So, the circle and the line have a single point of intersection and the line is perpendicular to the radius at that point.

Therefore, $3y - 4x + 15 = 0$ is a tangent to the circle with equation $x^2 + 2x + y^2 - 4x - 20 = 0$.

7 AOC and AOB are both isosceles as $AO = OC = OB$ all radii of circle.

Define $AOC = x$ and $AOB = y$

$OAC = \frac{180° - x}{2} = 90° - \frac{x}{2}$

and

$OAB = \frac{180° - y}{2} = 90° - \frac{y}{2}$ (base angles of isosceles triangles are equal)

Then,

$CAB = OAC + OAB$

$= \left(90° - \frac{x}{2}\right) + \left(90° - \frac{y}{2}\right)$

$= 180° - \frac{1}{2}(x + y)$

$BOC = 360° - x - y$ (angles around a point sum to 360°)

Therefore, CAB is half the size of BOC.

Answers

8 Define obtuse DOB to be x.

Then, $BCD = \frac{x}{2}$ (angle at the centre is double the angle at the circumference)

Reflex $DOB = 360° - x$ (angles around a point sum to 360°)

And $BAD = \frac{360° - x}{2} = 180° - \frac{x}{2}$ (angle at the centre is double the angle at the circumference)

Therefore, $BAD + BCD = 180° - \frac{x}{2} + \frac{x}{2} = 180°$, and opposite angles in a cyclic quadrilateral sum to 180°.

9 Define $DOC = x$

$DAC = DBC = \frac{x}{2}$ (angle at the centre is double the angle at the circumference)

So, $DAC = DBC$.

10 Proof shown

Chapter 10: Vectors

Maths challenge: 23

10.1 Position vectors

1 a $\begin{pmatrix} 9 \\ 16 \end{pmatrix}$ **b** $\begin{pmatrix} -4 \\ -8 \end{pmatrix}$ **c** $\begin{pmatrix} -1 \\ -1 \end{pmatrix}$

2 a $\begin{pmatrix} 0 \\ 24 \end{pmatrix}$ **b** $\begin{pmatrix} -2 \\ -16 \end{pmatrix}$ **c** $\begin{pmatrix} -3 \\ 39 \end{pmatrix}$

3 a $\begin{pmatrix} 14 \\ -6 \end{pmatrix}$ **b** $\begin{pmatrix} -28 \\ 12 \end{pmatrix}$ **c** $\begin{pmatrix} -18.5 \\ 7 \end{pmatrix}$

4 7.62

5 $2\sqrt{10}$

6 $3\sqrt{5}$

7 a 19.89

 b 80.54°

8 49 square units

9 11.59 square units

10 $\begin{pmatrix} -9 \\ 11 \end{pmatrix}$

11 $x = 4$ or -8

10.2 Solving geometric problems

1 a $-2\mathbf{a} - 9\mathbf{b}$

 b $3\mathbf{b} - 4\mathbf{a}$

 c $6\mathbf{a} + 6\mathbf{b}$

2 a $-\mathbf{a}$

 b $\mathbf{b} - \mathbf{a}$

 c $2\mathbf{a} - \mathbf{b}$

 d $2(\mathbf{b} - \mathbf{a})$

 e $2\mathbf{b}$

 f $2\mathbf{b} - \mathbf{a}$

3 a $\overrightarrow{AB} = \mathbf{b} - \mathbf{a}$

 b Proof showing that
 $\overrightarrow{NM} = \frac{1}{2}(\mathbf{b} - \mathbf{a})$
 so, $\overrightarrow{AB} = 2\overrightarrow{NM}$

4 a $\overrightarrow{BD} = \mathbf{e} - \mathbf{a}$

 b Proof showing that
 $\overrightarrow{OF} = \frac{2}{3}(\mathbf{e} + 2\mathbf{a})$ and $\overrightarrow{FC} = \frac{1}{3}(\mathbf{e} + 2\mathbf{a})$
 so, $\overrightarrow{OF} = 2\overrightarrow{FC}$

5 $ON:NB = 2:1$

6 $ON:NB = 2:1$

7 Proof showing that

 $\overrightarrow{PM}$ common
 $\overrightarrow{PN} = \overrightarrow{AM} = \mathbf{b}$
 $\overrightarrow{NM} = \overrightarrow{PA} = \mathbf{a}$

8 Proof showing that

 $\overrightarrow{AP} = \lambda(4\mathbf{c} - 2\mathbf{a})$
 $\overrightarrow{AP} = -\mathbf{a} + \mu\left(\frac{3}{5}\mathbf{a} + 4\mathbf{c}\right)$
 $\lambda = \mu$
 $-2\lambda = -1 + \frac{3}{5}\lambda$
 $\lambda = \frac{5}{13}$
 so, $\overrightarrow{AP} = \frac{5}{13}\overrightarrow{AC}$
 and $\overrightarrow{PC} = \frac{8}{13}\overrightarrow{AC}$

9 Proof showing that

 $\overrightarrow{CE} = \mu\left(-\frac{1}{2}\mathbf{a} + \mathbf{c}\right)$
 $\overrightarrow{CE} = \frac{2}{3}(-\mathbf{a} + \mathbf{c}) + \lambda\mathbf{a}$
 $\mu\mathbf{c} = \frac{2}{3}\mathbf{c}$
 $\mu = \frac{2}{3}$
 $-\frac{1}{2}\mu\mathbf{a} = -\frac{2}{3}\mathbf{a} + \lambda\mathbf{a}$
 $-\frac{2}{6} = -\frac{2}{3} + \lambda$
 $\lambda = \frac{1}{3}$

10 $\frac{1}{6}(2\mathbf{x} - \mathbf{y})$ or $\frac{1}{3}\mathbf{x} - \frac{1}{6}\mathbf{y}$

Answers

Mixed practice

1. **a** $4a^6c^{10}$
   b $27x^3y^{12}$
2. $\sqrt{375}$
3. $20x^4 - 16x^3 - 45x^2 + 36x$
4. **a** $2x + 3y - 12 = 0$
   b $3x - 2y - 6 = 0$
5. **a** $m^5 - m^3$
   b $48a^{10}b^{16}$
6. $n = -2$
7. 14
8. 3.93 cm^2
9. **a** 8.14 m
   b 290.3°
10. $\sqrt{40}$
11. Graph passing through 4, -6 and -3.5 on x-axis and -168 on the y-axis. Maximum in 2nd quadrant and minimum in 4th quadrant.
12. Correct counter example using, e.g., $m = -1$ and $n = -3$
13. **a** $p = 7$ and $q = 2$
    b
    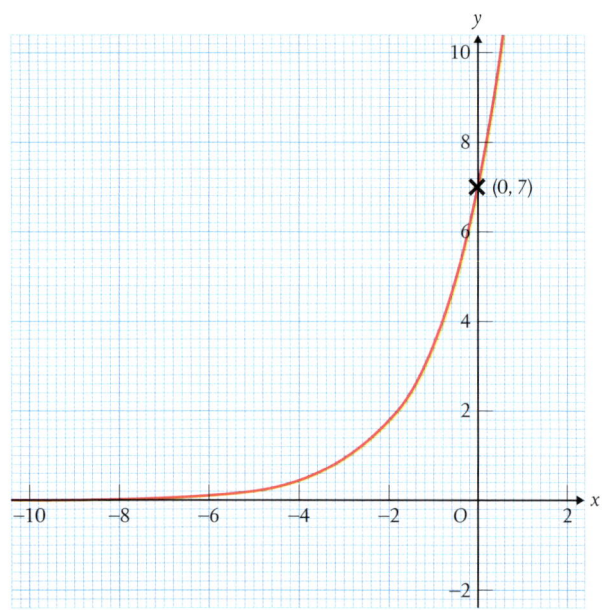
14. **a** $64 + 48x + 12x^2 + x^3$
    b 100 000
    c $16y^4 - 160y^3 + 600y^2 - 1000y + 625$
15. 61.5 cm (1 d.p.)
16. **a** $\frac{1}{55}$
    b $\frac{3}{5}$
17. **a** $4(x + 3)^2 - 41$ **b** minimum at $(-3, -41)$
18. 13 cm
19. $\left(\frac{4}{\sqrt{5}}, \frac{8}{\sqrt{5}} + 1\right) = \left(\frac{4}{\sqrt{5}}, \frac{8 + \sqrt{5}}{\sqrt{5}}\right) = \left(\frac{4\sqrt{5}}{5}, \frac{5 + 8\sqrt{5}}{5}\right)$
    $\left(-\frac{4}{\sqrt{5}}, -\frac{8}{\sqrt{5}} + 1\right) = \left(-\frac{4}{\sqrt{5}}, \frac{\sqrt{5} - 8\sqrt{5}}{\sqrt{5}}\right) = \left(-\frac{4\sqrt{5}}{5}, \frac{5 - 8\sqrt{5}}{5}\right)$

20. 65 or 18
21. 17.5°
22. **a**
    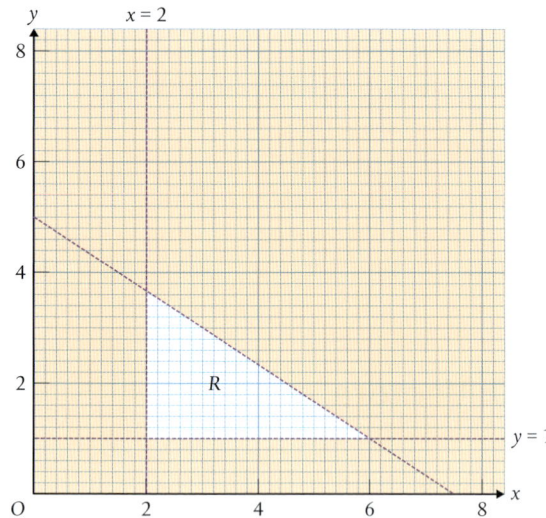
    b 22
23. $5 + 3\sqrt{2}$
24. $(7, 0), (2, 5)$
25.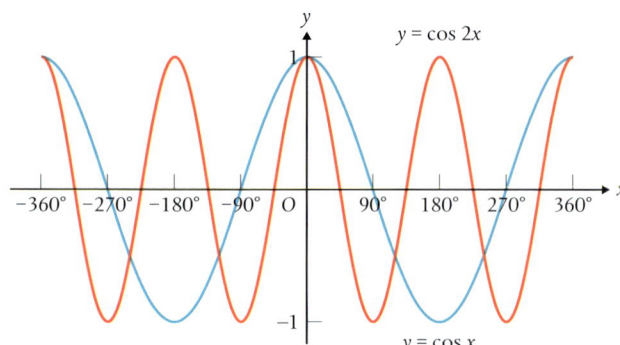
26. **a** $x \geq 4.5$
    b $-2.5 < x < 6$
    c $4.5 \leq x < 6$
27. $x = 0.4$
28. **a** $6\mathbf{b} - 2\mathbf{a}$
    b $6\mathbf{b} - 7\mathbf{a}$
    c $\frac{5}{13}$
29. 50.4°
30. 50 square units
31. **a** $3^3 - 6 \times 3^2 - 3 + 30 = 0$ or equivalent
    b $x = 3, x = 5, x = -2$
32. $-5(x - 2)^2 + 23$
33. **a** $(5, 12)$
    b $(10, 3)$
    c $(5, -4)$
    d $(8, 3)$

Answers

34 30°, 150°

35 $(y-5) = -\frac{4}{3}(x-7)$

36 $x = 0$ and $x = 3$

37 25

38 Shows e.g. $(7n+1)^2 = 7(7n^2 + 2n) + 1$ and $(7n-1)^2 = 7(7n^2 - 2n) + 1$

39 53

40 $y = x^2 + 12x + 42$

41 e.g. $(3n+3)^2 - (3n-3)^2 = 36n = 18 \times 2n$ so a multiple of 18

42 a $2 \times 2^3 - 9 \times 2^2 - 2 \times 2 + 24 = 16 - 36 - 4 + 24 = 0$

   b $(x-2)(2x+3)(x-4)$

43 180°

44 a $\frac{5}{3}$

   b 0.25

   c 162

   d $x^2 - 6x + 5$

   e $4 \times 4(-4)^3 + 24 \times (-4)^2 + 27 \times -4 - 20 = -256 + 384 - 108 - 20 = 0$

   f $-4, 0.5, -2.5$

45 Proof finding the point at which the line $4y + 3x = -14$ touches the circle $x^2 + 10x + y^2 + 12y + 36 = 0$ and then showing that the gradient of the radius at that point is perpendicular to the gradient of the line, since a tangent at a point on a circle has gradient perpendicular to the gradient of the radius at that point.

First rearrange $x^2 + 10x + y^2 + 12y + 36 = 0$ into the form $(x+5)^2 + (y+6)^2 = 25$ to show it is a circle with centre $(-5, -6)$ and radius 5.

Rearrange $4y + 3x = -14$ into the form $y = -\frac{3}{4}x - \frac{7}{2}$

Substitute $y = -\frac{3}{4}x - \frac{7}{2}$ into $(x+5)^2 + (y+6)^2 = 25$ to give
$\frac{25}{16}(x+2)(x+2) = 0$

So, $x = -2$ and the line intersects the circle at a single point.

When $x = -2$, $y = -\frac{3}{4} \times (-2) - \frac{7}{2} = \frac{3}{2} - \frac{7}{2} = -2$

Point of intersection is $(-2, -2)$.

Gradient of radius to $(-2, -2) = \frac{-2-(-6)}{-2-(-5)} = \frac{4}{3}$

Gradient of tangent $= -\frac{3}{4}$ which is equal to the gradient of the line.

Therefore, $4y + 3x = -14$ is a tangent to the circle with equation $x^2 + 10x + y^2 + 12y + 36 = 0$.

Index

adding and subtracting vectors 160, 161, 164
adding to and subtracting from functions 98
algebra
  completing the square 21
  expanding brackets 12–14
  factorising 17–19
  laws of indices 10
algebraic division 18–19, 20
algebraic fractions
  adding or subtracting 24–5
  multiplying or dividing 23–4
  simplifying 23–5
  solving equations 104
Archimedes 1
area of a triangle 131–2
area under a curve 77–8
asymptotes 72, 86

brackets, expanding 12–14

circles, graphs of 68–9
coefficients 13–14
colinear vectors 164
common factors 17–18
complement of a set 137
completing the square 21, 107–8
composite functions 91–2
conditional probability 144–5
congruence 167
consecutive numbers, expression for 149
cosine function (cos) 126, 127
  exact values 48
  graph 48
  see also trigonometric equations; trigonometry
cosine rule 131
counter example, disproof by 153
cubic graphs 40–2
curved graphs 77–8

deduction, proof by 149
denominators, rationalising 6–7
difference of two squares 17
discriminant 35
disproof by counter example 153
distance between position vectors 160, 161
dividing indices 2
dividing polynomials 17–18
domain of a function 84–6, 87

elimination method, simultaneous equations 111

equations
  with algebraic fractions 104
  quadratic 33–5, 107–8
  simultaneous 111–12
  trigonometric 104
estimated probability 137
even numbers, expression for 149
exhaustion, proof by 153
expanding brackets 12–14
  with surds 4, 5
experimental probability 137
exponential graphs 72–3
exponential growth and decay 72

factor theorem 18
factorising 17–19
feasible region of a graph 117–18
fractions
  algebraic 23–5, 104
  rationalising denominators 6–7
functions 84–6
  composite 91–2
  inverse 95
  transforming 98

geometric proof 155–6
gradient of a curve 77
gradient of a linear graph 29–30
graphs
  of circles 68–9
  cubic 40–2
  exponential and reciprocal 72–3
  of inequalities 117–20
  linear 29–32
  non-linear 77–8
  quadratic 33–6
  quartic 43
  stretching 62–3, 64, 65
  translation and reflection 54–6, 59, 65
  trigonometric 48–51

independent events 137
indices, laws of 2
  simplifying algebraic expressions 10
inequalities 117–20
integers ($\mathbb{Z}$) 84
intercepts
  linear graphs 29
  quadratic graphs 37
intersection of sets (∩) 137, 139
inverse functions 95
irrational numbers 84

laws of indices 2
  simplifying algebraic expressions 10
linear graphs 29–32
linear programming 117–20

many-to-one relationships 84
mapping diagrams 84
maxima and minima 33, 36, 89
multiples of vectors 164
multiplicative law for independent events 137
multiplying functions by a scalar 98
multiplying indices 2
multiplying position vectors by a scalar 160, 161
mutually exclusive events 137

negative indices 2
negative numbers in inequalities 120

objective functions 118
odd numbers, expression for 149
one-to-many relationships 84
one-to-one relationships 84

parabolas 33
  see also quadratic graphs
parallel lines 30, 31
parallel vectors 164–5
Pascal's triangle 13–14
perfect squares 21
period of a graph 48
perpendicular lines 30, 32
polynomials, factorising 18–19
position vectors 160–1
  resultant 164
  solving geometric problems 164–5
powers of indices 2
probability 137–9
  conditional 144–5
  P(A and B) 141
  P(A or B) 141
  P(not A) 140
  tree diagrams 141, 144
proof
  by deduction 149
  disproof by counter example 153
  by exhaustion 149
  geometric 155–6
Pythagoras' theorem 126
  proof of 150

quadratic equations 107–8
  graphical solution 33–4
quadratic expressions
  completing the square 21
  factorising 17
quadratic formula 34–5, 107
quadratic graphs 33–6
quartic equations 110
quartic graphs 43

range of a function 84–6
rational numbers 84
rationalising denominators 6–7
real numbers ($\mathbb{R}$) 84
reciprocal graphs 72–3, 86
reflection 54, 56, 59, 65
relations 84
resultant vectors 164
roots 2
  see also surds
roots of a cubic function 40–1
roots of a quadratic function 33–5, 107

scale factors 62–3, 64, 65
self-inverse functions 95
set notation 137
simultaneous equations 111–12
sine function (sin) 110, 126
  exact values 48
  graph 48–50
  see also trigonometric equations; trigonometry
sine rule 131
sketch graphs 29, 37
straight line, equation of 29–32
stretching graphs 62–3, 64, 65
substitution method, simultaneous equations 111–12
surds 4
  rationalising denominators 6–7

tangent function (tan) 72, 126
  exact values 48
  graph 48, 50–1
  see also trigonometric equations; trigonometry
tangent to a curve 77
transforming functions 98
translation 54–5, 59, 65
  graphs of circles 68, 69
  trigonometric graphs 63
trapezium rule 77–8
tree diagrams 141, 144
trigonometric equations 104
trigonometric graphs 48–51
trigonometry 126–7
  area of a triangle 131–2
  sine and cosine rules 131
turning points 33, 36

union of sets (∪) 137

vectors 160–1
  solving geometric problems 164–5
Venn diagrams 137, 139

y-intercept 29, 37, 40

$\mathbb{Z}$ (integers) 84